T0395418

Surface Science and Synchrotron Radiation

Online at: https://doi.org/10.1088/978-0-7503-3847-9

Surface Science and Synchrotron Radiation

Phil Woodruff
Physics Department, University of Warwick, Coventry, UK

IOP Publishing, Bristol, UK

ISBN 978-0-7503-3847-9 (ebook)
ISBN 978-0-7503-3845-5 (print)
ISBN 978-0-7503-3848-6 (myPrint)
ISBN 978-0-7503-3846-2 (mobi)

DOI 10.1088/978-0-7503-3847-9

Version: 20231201

IOP ebooks

British Library Cataloguing-in-Publication Data: A catalogue record for this book is available from the British Library.

Published by IOP Publishing, wholly owned by The Institute of Physics, London

IOP Publishing, No.2 The Distillery, Glassfields, Avon Street, Bristol, BS2 0GR, UK

US Office: IOP Publishing, Inc., 190 North Independence Mall West, Suite 601, Philadelphia, PA 19106, USA

Contents

Preface

The start of my research career in the late 1960s coincided with the introduction of commercial, demountable, stainless steel ultra-high-vacuum systems, making it possible to add and remove different surface probes with relative ease (relative, that is, to the era of glass systems that relied on skilled glassblowers for building any experimental system). The result was the rapid development of new surface science techniques involving electron, photon and ion impacts. The use of this range of complementary methods made it possible to gain a better understanding of the structural, electronic and chemical properties of surfaces and their interrelationships. It was possible to establish a range of these techniques in a researcher's home laboratory.

The creation and development of synchrotron radiation sources, initially in the late 1970s, made possible a range of essentially new experimental techniques to determine the properties, particularly the structural properties, of a wide range of physical and biological materials. However, performing experiments at centralised, largely national, synchrotron radiation facilities takes a researcher from the comfort of his or her own laboratory. Is it worth the effort? For many researchers the answer is very clearly yes. The information that can be gained is often not available in the home laboratory, or is of a vastly superior quality. Some researchers focus their attention almost exclusively on synchrotron radiation methods, but many simply use synchrotron radiation experiments to complement their work in their home laboratory.

So, what is to be gained for a surface scientist in making the effort to use synchrotron radiation? The fact that surface science requires ultra-high vacuum, that a surface needs to be characterised *in situ* by several complementary methods, and that synchrotron radiation surface science experiments must be performed away from the home laboratory makes them challenging; but I hope that in this book I have shown the effort is worthwhile. Not only does synchrotron radiation offer significant advantages in the quality of data obtainable from otherwise established standard home laboratory methods, such as photoemission, but it offers entirely new techniques that have no equivalent in the home laboratory.

Author biography

Phil Woodruff

Phil Woodruff is Emeritus Professor of Physics at the University of Warwick, where he has been based for more than 50 years, albeit with shorter periods spent at Bell Telephone Laboratories in Murray Hill (NJ, USA) in the early 1980s and a visiting position at the Fritz Haber Institute in Berlin from 1999 to 2011. His research in experimental surface science has led to the award of a number of national and international prizes in the UK, Germany and the USA, and he was elected a Fellow of the Royal Society in 2006.

IOP Publishing

Surface Science and Synchrotron Radiation

Phil Woodruff

Chapter 1

Why use synchrotron radiation to study surfaces?

The development of modern surface science in the latter part of the 20th century, and beyond, has highlighted the need for the application of a range of complementary techniques to gain a proper understanding of the structural, electronic and chemical properties of the outermost few atomic layers of solid surfaces. Initially, these techniques relied on the use of special ultra-high-vacuum (UHV) instrumentation that can be operated in a typical university or industrial 'home' laboratory. However, the creation of centralised, typically national, facilities providing sources of synchrotron radiation has opened up new important capabilities for surface science, in some cases enhancing the established home laboratory techniques, but also introducing new methods with no direct equivalent achievable in a home laboratory. This initial chapter outlines the complementarity of these home and centralised capabilities.

1.1 'Modern' surface science

Experimental studies of the electronic and chemical properties of solid surfaces ('surface physics' and 'surface chemistry') were already being performed in many different laboratories in the early part of the 20th century, and while the applications of these two types of investigation were quite different, many of the pioneers saw no distinction in the fundamental physics underpinning the science. When Irving Langmuir was awarded the Nobel Prize in Chemistry in 1932 for his work on surface chemistry, he was employed by the General Electric Company and working on incandescent light bulbs, so he was also a 'surface physicist'. The more general discipline of 'surface science' is clearly not new! However, the introduction of stainless steel UHV chambers and ion pumps in the 1960s transformed the subject and led to a huge increase in activity. Prior to this, vacuum vessels were made of glass and outgassing of the inner surfaces to allow ultra-high vacuum to be achieved required baking to 400 °C or more. Pumping was typically achieved by mercury

doi:10.1088/978-0-7503-3847-9ch1

diffusion pumps fitted with liquid nitrogen traps. Making any change to the experiment conducted in such chambers required access to skilled glassblowers who could cut open and re-join the chamber components. By contrast, with stainless steel systems baking temperatures are typically little more than 150 °C, but most importantly a sample in the chamber can be accessed by simply unbolting a metal flange, while a wide variety of components for mechanical manipulation and to introduce probes such as electron guns, ion guns and X-ray sources, together with suitable detectors, can be mounted directly onto these easily changeable flanges. The result of this transformation in the underlying technology led not only to a large increase in the number of surface science researchers, but also to a very significant increase in the number of techniques that came to be applied to study surfaces, leading to a new phase of 'modern' surface science. A subset of these is now in routine use by many groups worldwide in laboratories in universities, research institutes and some companies. Detailed descriptions of these methods and their applications can be found in several other books such as Woodruff (2016) and Kolasinski (2019).

During the period of the exploration of alternative new techniques in the 1970s and 1980s, the tool of synchrotron radiation started to become available for applications in a wide range of different fields in the physical and biological sciences as well as in engineering. New purpose-built sources of synchrotron radiation were brought into user operation in the 1980s, and increasingly sophisticated new sources are still being commissioned. Synchrotron radiation provides electromagnetic radiation that is readily 'tuneable' over the continuous range of photon energies from the far-infrared to hard X-rays (meV to hundreds of keV), with variable polarisation and extremely high spectral brightness, which means that the photon flux deliverable to a small sample is very many orders of magnitude larger than that possible using standard 'laboratory' sources. The availability of this radiation has had two distinct types of impact on surface science. First, it greatly extends the capabilities of existing standard laboratory techniques based on incident photons (most notably different variants of the basic photoemission technique); but, second, it offers entirely different techniques for which there is no practical laboratory source equivalent. It is these new techniques, and the advances in existing techniques and their applications, which are the subject of this book. Prior to this core material covering the techniques, chapter 2 provides a more detailed description of the properties of synchrotron radiation itself and the means of delivering it to a surface science sample. In this short introductory chapter some of the underpinning principles of all surface science techniques are summarised, together with their particular reference to methods exploiting synchrotron radiation.

One further important and somewhat more recent development in the broad area of surface science concerning the growing interest in nanoscience—understanding the physics and chemistry of materials physically constrained in one, two or three dimensions to the nanometre scale. Of course, the traditional 'surface' of surface science is such a one-dimensionally constrained material, corresponding to the outermost few atomic layers with a thickness of ~1 nm. Ultra-thin films of metals and semiconductors have also been subjects of surface science studies for some years.

Moreover, it had long been recognised that in heterogeneous catalysis the size of the catalyst particles, down to the nanometre range, can play an important role in determining their efficacy and selectivity. In addition, the idea of (naturally occurring) 'layer materials' comprising atomic layers with strong coupling within the layers but weak coupling between them, such as graphite and transition-metal dichalcogenides, has long been a subject of interest. However, true single-atomic-layer materials such as graphene and artificially created layer metamaterials are new, and their properties and applications have become a rich source of research. The tools of surface science and the availability of synchrotron radiation have also impacted this field, and will be illustrated here by a few examples.

1.2 Surface sensitivity and surface specificity

Understanding the physics and chemistry of solid surfaces generally requires techniques that can not only be *sensitive* to the small amount of material that lies within the outermost few atomic layers of the solid, but also provide information that is *specific* to these few atomic layers. Deeper below the surface there are generally very many more layers of bulk solid, and if there is no surface specificity in the probing technique, the weak signal from the surface will be swamped by the signal from the underlying bulk.

All the techniques to be discussed in this book are (of course) concerned with photon incidence, in the broad spectral energy range from the vacuum ultraviolet (or VUV, with energies of up to a few tens of eV) through soft X-rays (SXRs, hundreds of eV to a few keV) and on to hard X-ray energies (up to a few tens of keV). At the lower energies of this broad range, in particular, most solids are strongly absorbing, so an obvious question is, 'Does this lead to intrinsic surface specificity?'. As an example, figure 1.1 shows the attenuation length of copper (Cu) for normal incidence

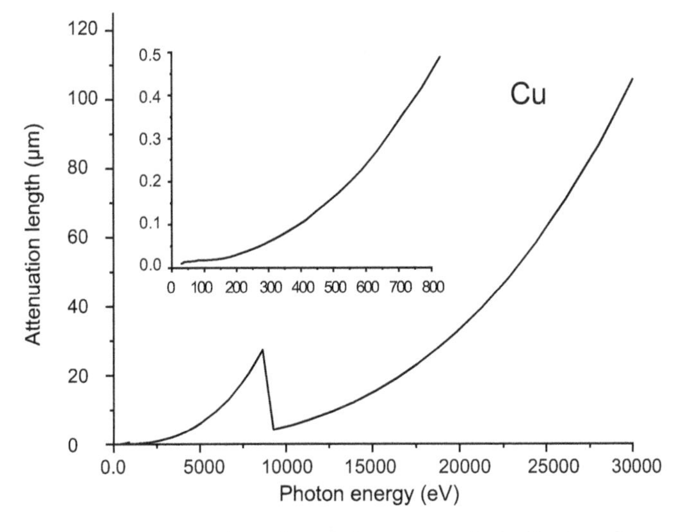

Figure 1.1. Attenuation length due to absorption of X-rays normally incident on Cu as a function of the photon energy. Data based on the updated web version of the original tabulations by Henke *et al* (1993).

of photons of energies covering this complete range, plotted from data updated from the original tabulations by Henke *et al* (1993).[1] The absorption leading to the attenuation of the photon beam as it passes through a material is dominated by photoionisation, and the sharp drop in the attenuation length for Cu around 8000 eV corresponds to the threshold for photoionisation of the 1s core level of Cu; this can easily be resolved experimentally much more precisely than indicated by the figure, but is smoothed out by the rather coarse sampling in energy of this data set.

In the hard X-ray range, the attenuation length is ~tens of microns, so the region of illumination is dominated by the bulk, not the surface. However, even at the lowest energies of ~100 eV, the attenuation length is ~170 nm, very much larger than the thickness of the outermost few atomic layers of a surface of ~1 nm, so at no energy in this very broad range does the penetration depth of the incident X-rays provide surface specificity under the condition of normal incidence. Of course, grazing incidence can reduce the attenuation *depth* by a factor $1/\cos\theta$, where θ is the angle of incidence relative to the surface normal, but even at the lowest energies a very small grazing incidence angle would be necessary to achieve true surface specificity. For example, to reduce the attenuation depth to ~1 nm at 100 eV from an attenuation length value of ~170 nm at normal incidence would require an incidence angle of 89.66°, i.e., a grazing incidence angle of 0.34°. At higher energies much smaller grazing angles would be required to ensure that photoabsorption would lead to true surface specificity.

However, the grazing incidence of X-rays can also lead to reduced interaction with the underlying bulk if this angle is less that that corresponding to the condition for total reflection. In conventional visible-light optics the materials used to fabricate optical components (lenses, prisms, etc.) are fabricated from materials that have a refractive index significantly greater than unity (e.g., in the range ~1.3–1.5). In these materials grazing incidence at the glass–vacuum (or glass–air) interface from inside the material can lead to *total internal reflection*. The condition for this can readily be obtained from Snell's law:

$$n_1 \sin \theta_{inc} = n_2 \sin \theta_{refr},$$

where θ_{Inc} and θ_{refr} are the angles of incidence (from a medium with refractive index n_1) and refraction (into a medium with refractive index n_2), defined relative to the interface normal. In the case of total internal reflection $n_2 = 1$ (the value for vacuum and almost the value for air), so if $n_1 > 1$ the sine of the refracted angle reaches its maximum possible value of unity when $\theta_{refr} = 90°$, i.e., the refracted beam lies parallel to the interface. There is thus no refracted (transmitted) beam for any incidence angle greater than a critical angle θ_c, defined as

$$\theta_c = \sin^{-1}(n_2/n_1).$$

[1] Available online at http://henke.lbl.gov/optical_constants/atten2.html.

At X-ray photon energies, however, the refractive index of vacuum is still unity (by definition), but the refractive index of solids is slightly less that unity; under these conditions, this total reflection condition arises for external rather than internal reflections. The refractive index of a material at X-ray energies can be related to the atomic scattering factors of its constituent atoms. The refractive index at X-ray energies is generally written as

$$n = 1 - \delta + i\beta,$$

where δ is the deviation in the real part of the refractive index from unity, which may be only $\sim 10^{-6}$ at X-ray energies, and β, the imaginary component, accounts for absorption. This refractive index can be related to the forward (zero scattering angle) atomic scattering factor of the radiation of the atoms comprising the solid, $f_r^0 - if_i^0$, where the subscripts r and i denote the real and imaginary components:

$$n = 1 - \frac{n_a r_e \lambda^2}{2\pi}[f_r^0 - if_i^0],$$

where n_a is the atomic density, λ is the wavelength and r_e is the classical electron radius:

$$r_e = \frac{e^2}{4\pi\varepsilon_0 mc^2},$$

with e being the electron charge, ε_0 the permittivity of free space, m the electron mass and c the speed of light. If we can assume the absorption is very weak (i.e., assume that $\beta = 0$), then

$$\sin\theta_c = 1 - \delta.$$

Evidently, this critical value of the angle of incidence is quite close to 90°, i.e., the incidence is grazing to the surface. It is therefore more useful to discuss the critical reflectivity conditions in terms of the *grazing* angle of incidence, ϕ_i, which is the complement of θ_i, so

$$\cos\phi_c = 1 - \delta,$$

and as this grazing angle is small, expressing $\cos\theta_c$ by its series expansion and taking only the leading terms leads to

$$\phi_c \simeq \sqrt{2\delta}.$$

Taking again the case of copper, but now assuming a 'hard' X-ray wavelength of 1 Å (1 Å = 0.1 nm)[2] the value of δ is 11.05×10^{-6}, leading to a value of the critical grazing angle of 4.7 mrad or 0.27°. Total reflection, which occurs for all grazing incidence angles less than this critical value, would seem to imply no penetration of the surface and thus complete surface specificity. However, by considering the

[2] The ångström unit (1 Å = 0.1 nm) is rather convenient in discussing X-ray diffraction and the structure of solids, not only because the wavelengths of interest are ~1 Å, but also because typical nearest-neighbour interatomic distances typically fall in the range ~2–4 Å, so this unit will be used frequently in this book.

matching at the interface of waves on either side of the interface, it transpires that there is a refracted wave at the interface but its wavevector is complex. In particular, its wavevector perpendicular to the surface is imaginary, thereby corresponding to an evanescent wave, the amplitude of which decays exponentially into the material from the surface. This is an optical analogue of the situation of an electron impinging on a surface at an energy corresponding to a band gap in the solid, which can still couple to an electronic surface state in which an electron can travel parallel to the surface but has an exponentially decaying amplitude perpendicular to the surface. The existence of this evanescent wave means that the incident X-rays can interact with sub-surface atoms to a depth determined by the decay length of this wave, d, given by

$$d = \frac{\lambda}{4\pi\sqrt{\sin^2\theta - n^2}} = \frac{\lambda}{4\pi\sqrt{\sin^2\theta - \sin^2\theta_c}}.$$

Re-expressing these equations in terms of the grazing incidence angle ϕ_{Inc} ($\cos\theta_{\mathrm{inc}} = \sin\phi_{\mathrm{inc}}$), using the identity $\cos^2\phi = 1 - \sin^2\phi$ and noting that for small values of ϕ, $\sin\phi \cong \phi$ leads to

$$d = \frac{\lambda}{4\pi}\frac{1}{\sqrt{\phi_c^2 - \phi_{\mathrm{inc}}^2}}.$$

For example, if one assumes an incident grazing angle of one half of the critical value (~5 mrad) for total reflection from Cu, one obtains a value for d of approximately 2 nm. Evidently, this does not guarantee true surface specificity, although the greatly reduced sampling depth should reduce the substrate scattering signal relative to that from the surface. Notice that the attenuation length in Cu (figure 1.1) due to absorption at this photon energy is 6.4 μm, so at this very grazing incidence angle of 0.135° the attenuation depth is only ~15 nm, thus in this case elastic scattering at this grazing angle is significantly more effective than inelastic scattering in determining the degree of surface specificity. Of course, the wavelength of 1 Å corresponds to an energy of ~12 keV, which, as shown in figure 1.1, corresponds to quite a low attenuation length due to being only a few keV above the Cu 1s photoionisation threshold. Notice that at no grazing angle does the elastic scattering give total surface specificity. The minimum possible value of d is given by $\theta_{\mathrm{Inc}} = 0$ when $d = \frac{\lambda}{4\pi\phi_c}$, which in this example corresponds to a value of 1.7 nm.

The overall conclusion, therefore, is that photon-in/photon-out X-ray techniques, such as X-ray diffraction, cannot be fully surface specific due to either inelastic or elastic scattering, although grazing incidence can very significantly enhance the degree of surface specificity.

However, while photon incidence does not intrinsically provide true surface specificity, a number of surface techniques involve measuring the emission of electrons induced by photoabsorption, most obviously in photoemission, but also techniques based on studies of photoabsorption, such as X-ray absorption near edge structure (XANES), can also be detected via the emission of secondary electrons (see chapter 3.6). Photoemission involves the measurement of the energy spectrum of the electrons emitted as a result of the incident photons, and one measures experimentally the intensity and the spectral peak shape of photoelectrons that retain their characteristic kinetic energy

$$E_{\text{kin}} = h\nu - E_b,$$

where $h\nu$ is the photon energy and E_b is the initial-state binding energy (both E_b and E_{kin} being relative to the same reference energy—usually the Fermi level of the solid). In this experiment the surface specificity is determined not by the long incident photon attenuation length, but by the much shorter attenuation length of the escaping electrons. This arises from the short inelastic scattering mean free path for electrons in the energy range relevant to photoemission studies, generally ~10 eV to a few keV, although the effective attenuation length (or EAL) is also influenced by elastic scattering, as described below. The dominant mechanism of inelastic scattering in most of this energy range is plasmon scattering (the creation of quantised longitudinal plasma oscillations of the valence electrons of the solid), although at higher energies core-level ionisation plays a role, whereas at the lowest energies, below the threshold to plasmon excitation (typically ~25–30 eV) only electron–hole excitations of the valence states are possible. Figure 1.2 shows the results of experimental determinations of the attenuation length of electrons escaping from a wide range of different solid materials, measured by Seah and Dench (1979). These measurements were performed by measuring the attenuation of core-level photoemission signals from a bulk solid coated with ultra-thin films of different materials as a function of the film thickness d. The measured intensity, I, is related to the intensity with no covering thin film, I_0, by

$$I = I_0 e^{(-d/\lambda \cos\theta)},$$

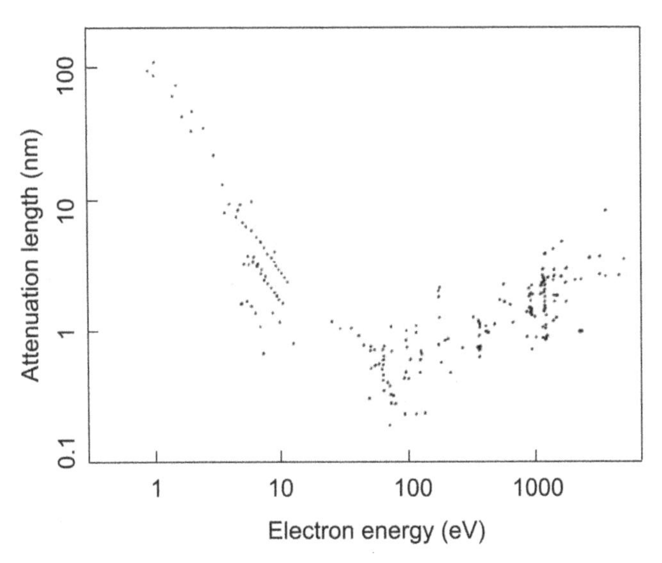

Figure 1.2. Measured values of the attenuation length of electrons passing through a range of different solid materials as a function of their kinetic energy. Reproduced from Seah and Dench (1979) John Wiley & Sons.

where λ is the attenuation length and θ is the angle of collection relative to the surface normal. Notice that this attenuation length is not equivalent to the inelastic scattering mean free path due to the role of elastic scattering by the atoms in the overlayer film. Of course, elastic scattering is a conservative process that leads to no attenuation, but elastic scattering (at angles greater than 0°) increases the length of the path that the electrons must travel through the film in order to escape; this longer ('zig-zag') path increases the probability of inelastic scattering, leading to additional attenuation relative to that expected ($d/\cos\theta$) for a direct path length to the surface. In practice, this role of elastic scattering is of comparable importance in determining the attenuation length, which can be ~50% of the inelastic scattering mean free path. Figure 1.2 shows clearly that for emitted electrons in the kinetic energy range from ~10 to 1000 eV, the attenuation length is ~1 nm or less, a distance corresponding to ~3–5 atomic layers, the region of interest in surface science. Inelastic scattering (combined with elastic scattering) does, therefore, ensure that techniques which rely on detecting emitted electrons in this energy range that have retained their initial kinetic energy of their excitation are surface specific. This is the basis of the surface specificity of standard surface analysis techniques such as X-ray photoelectron spectroscopy (XPS) and Auger electron spectroscopy (AES). Notice, though, that the increase in attenuation length at higher energies does mean that data from shallowly buried interfaces becomes accessible. The attenuation length is commonly regarded as scaling as $\sim E_{kin}^{1/2}$, although the dependence seems to be stronger than this, as described in section 3.3. This is the basis of the technique of hard X-ray photoelectron spectroscopy (HAXPES), described in chapter 3.

Of course, there are other methods of achieving some degree of surface specificity. For example, in techniques that provide element-specific information, surface specificity is achieved if the detected signals come from elemental species that are only present in or on the surface, such as adsorbed atoms or molecules. This is true even if the method of detection, such as in X-ray emission, is not inherently surface specific. Other methods, such as detecting X-ray diffraction signals specific to the periodicity of the surface but not the bulk, are key to achieving true surface specificity in surface X-ray diffraction (SXRD), as is discussed in detail in chapter 4.

1.3 Complementary techniques

It was well-established in the early development of the large range of 'modern' surface science techniques that no one method can fully characterise the properties of a surface. A minimum requirement is the use of complementary methods to determine the composition, structure and electronic properties and, particularly for molecular systems, the vibrational properties. The same is true of methods based on the use of synchrotron radiation, and indeed the convenience of conducting experiments that can be performed in a researcher's home laboratory clearly means that these techniques form the bedrock of any investigation. Complete descriptions of these techniques are readily accessible in other books, as mentioned in section 1.1 above, and will not be described in detail here. Performing surface science studies using synchrotron radiation can be challenging. The allocated time for a synchrotron radiation surface science experiment is rarely in excess on one week or so, and the frequent demand for time-consuming *in situ* sample preparation means that not all of the allocated beamtime can be spent taking data. Performing synchrotron radiation surface science experiments is therefore only appropriate if the information to be gained cannot be obtained through the use of techniques that do not necessitate synchrotron radiation. It is therefore appropriate to consider a brief discussion of the relative strengths and limitations of standard laboratory techniques relative to those using synchrotron radiation; more detailed comparisons are included in the later chapters.

In the case of surface composition, the standard laboratory techniques of XPS and AES, in some cases aided by ion-scattering techniques, provide effective methods to determine the surface elemental composition, although synchrotron radiation makes it possible to obtain spatially resolved compositional information with significantly better resolution using scanning photoemission microscopy (SPEM) and imaging photoelectron emission microscopy (PEEM). Determining the elemental composition at (shallowly) buried interfaces using XPS and AES is limited by the attenuation length of the escaping electrons, particularly using Al or Mg Kα radiation, which limits the electron kinetic energies to below 1500 eV (and mostly to lower energies). Using higher photon energies from synchrotron radiation in the HAXPES technique can allow somewhat deeper probing of the elemental composition. Non-synchrotron radiation techniques based on ion scattering (notably, for high depth resolution, medium-energy ion scattering, MEIS, using ~100–200 keV H^+ or He^+ ions) are well-suited to providing this information, but this involves specialised instrumentation that is not widely accessible. However, if one wishes to determine the *chemical* composition of shallowly buried interfaces, i.e., not only the elemental species but also their chemical state, then synchrotron-radiation-based HAXPES is perhaps the only technique available.

The benchmark technique for determining the structure of surfaces with good long-range order is low-energy electron diffraction (LEED). The basic instrumentation—a LEED 'optics'—is installed in vast numbers of UHV surface systems worldwide, and is the basic instrumentation required to measure both the diffraction pattern (to determine the surface periodicity) and also the diffracted beam intensities (to determine atomic coordinates). The main limitation of LEED for quantitative structure determination is that the high cross-section of atoms for elastic scattering of electrons in the energy range used (~50–400 eV) means that multiple scattering is very important, so structure determination can only be achieved by comparing the experimentally measured diffracted intensities with simulated ones based on multiple scattering calculations for a succession of trial structures until a good fit is obtained. The technique works well for small unit mesh structures but becomes computationally very demanding for large unit mesh structures. This problem is exacerbated by the fact that large unit mesh structures typically have a large number of structural variables (coordinates of a large number of atoms within the unit mesh). The standard LEED technique is only applicable to surfaces with long-range order. Laboratory techniques based on ion scattering can also provide some quantitative structural information. The most obvious direct competitive technique requiring the intense, highly collimated X-ray beam only available through synchrotron radiation is SXRD. It is based on the same physical principles as LEED, but the weaker scattering means that data simulation can be performed using much less computationally demanding single-scattering simulations. This is a particular advantage for large unit mesh structures. In addition, however, the wide-energy-range tuneability of synchrotron radiation in both soft and hard X-ray regions gives access to quite different structural techniques, namely extended and near-edge X-ray absorption fine structure (EXAFS and NEXAFS) and photoelectron diffraction, which are based on the coherent interference of photoelectron wavefields. These techniques provide element-specific local structural information on surfaces that need not have long-range order.

The primary probe of surface electronic structure, photoemission from both core and valence electronic states, can be performed with both standard laboratory sources and synchrotron radiation. However, the fixed energies of the standard laboratory sources, 21.2 and 40.8 eV (He I and He II emission lines) for valence-band ultraviolet photoelectron spectroscopy (UPS) and 1253.6 and 1486.6 eV (Mg K_α and Al K_α) are significantly constraining. For example, the mapping of electronic bands' valence using angle-resolved photoelectron spectroscopy (ARPES) requires the ability to vary the photon energy more freely in order to distinguish two-dimensional and three-dimensional band states. The availability from synchrotron radiation of highly focussed beams of very significantly enhanced intensities also make possible ARPES measurements at much higher resolution in energy and momentum transfer. Synchrotron radiation also makes possible higher-spectral-resolution measurements of core-level photoemission and the associated 'chemical shifts' that distinguish atoms of the same element in distinct electronic environments. However, in addition to these enhancements of standard UPS, ARPES and XPS investigations, the ability to freely vary the photon energy in the SXR range makes entirely new experiments, such as investigations of resonant photoemission and the core-hole clock technique, possible.

Characterisation of vibrational states at surfaces are generally well-served by the standard laboratory techniques of high-resolution electron energy-loss spectroscopy (HREELS) and reflection–absorption infrared spectroscopy (RAIRS); RAIRS offers much better spectral resolution and the ability to study surfaces in the presence of significant pressures of reactant gases, while HREELS allows one to access additional vibrational modes that have no dynamic dipole moment perpendicular to the surface. However, the limitations of standard blackbody ('globar') sources in the far-infrared means that standard RAIRS experiments cannot access the low-frequency modes typically due to adsorbed molecules moving against the surface. Synchrotron radiation proves to be a much better source of far-infrared radiation for these experiments, but these experiments have nevertheless proved challenging and have fallen out of favour. The enhanced resolution achievable in XPS also has been shown to allow element-specific vibrational modes to be detected.

Real-space imaging of surfaces has been a growing trend in surface science for several decades, and synchrotron radiation offers the possibility of photoemission and photoabsorption imaging with resolution down to a few tens of nanometres or better. These methods have had particularly strong impact in understanding the electronic and magnetic properties of surfaces. Nevertheless, synchrotron radiation cannot offer any technique providing true atomic-scale spatial resolution of surfaces akin to that provided by scanning tunnelling microscopy (STM) and atomic force microscopy (or AFM). For surfaces with good long-range order these techniques provide the same quantitative information (in real space) provided by LEED patterns (in reciprocal space), namely, the periodicity of the associated unit mesh. These techniques also provide some insight into the likely general location of atoms and molecules within the unit mesh, although they do not, in general, provide reliable lateral dimensions or relative heights of these atoms within the unit mesh; in some cases, this information can be extracted by reliable theoretical modelling. Most significantly, however, these techniques provide direct real-space information on the *disorder* in the surface, such as the coexistence of different locally ordered phases or the influence of specific surface defects, such as atomic steps, on local order. This information is not available in other ways and provides potentially very important characterisation of a sample that may help the interpretation of the results of other, more spatially averaging techniques both with and without synchrotron radiation. Scanning tunnelling spectroscopy (or STS) can also provide unique information on local electronic and vibrational structure, although these extensions of the basic STM technique are exceptionally challenging experimentally.

1.4 What's special about synchrotron radiation?

Much fuller details of the characteristics of synchrotron radiation both at the source and at the sample are provided in chapter 2. In this section, only a brief description of these properties are summarised to highlight why an experimentalist may deem it helpful, or even essential, to leave the comfort of his or her home laboratory to run experiments at a synchrotron radiation facility. This is certainly an important question. Modern surface science techniques are well-established but are still not exactly routine; working in a UHV environment can be challenging and time consuming. There need to be good reasons to choose to combine these problems with the need to perform complete experiments with *in situ* sample preparation in a laboratory removed from home with the tight time constraints of allocations of synchrotron radiation beamtime—rarely more than a few days. The scientific benefits need to be significant. The key properties of synchrotron radiation that may offer these benefits are as follows:

- Broadband radiation from the far-infrared to hard X-rays allowing widely and continuously tuneable delivery to a sample over energy ranges up to a factor of ~10 at a single experimental station within this broad range. Additional sources with more harmonic character.
- The radiation is pulsed, but with typical pulse lengths in the picosecond range and a frequency ~500 MHz it can be regarded as quasi-continuous for most users.

- The radiation is emitted in an extremely narrow beam. While the angular width is smallest at the highest photon energies, the characteristic width is typically ~0.2 mrad (0.01°). It is therefore straightforward to deliver a high photon flux to a small sample many metres from the source.
- The extremely low photon emittance of the radiation beams (the product of the beam size and its divergence) means that they can be focussed to submicron dimensions (or in some cases tens of nanometres) at a sample.
- At the most modern sources one can exploit a high degree of coherence in the radiation.
- Characteristically, the radiation is linearly polarised, but circularly and elliptically polarised beams can be achieved, as can rotation of the plane of polarisation of linearly polarised radiation.

References

Henke B L, Gullikson E M and Davis J C 1993 X-ray interactions: photoabsorption, scattering, transmission, and reflection at E = 50–30,000 eV, Z = 1–92 *At. Dat Nucl. Data Tables* **54** 181–342

Kolasinski K W 2019 *Surface Science: Foundations of Catalysis and Nanoscience* 4th (New York: Wiley)

Seah M P and Dench W A 1979 Quantitative electron spectroscopy of surfaces: a standard data base for electron inelastic mean free paths in solids *Surf. Interface Anal.* **1** 2–11

Woodruff D P 2016 *Modern Techniques of Surface Science* (Cambridge: Cambridge University Press) 3rd edn

Chapter 2

Synchrotron radiation: at the source and at the sample

In this chapter, the basic properties of synchrotron radiation are introduced, initially from the historical perspective of charged particle accelerators, distinguishing cyclotron radiation from synchrotron radiation. Following a more detailed explanation of the characteristics of bending magnet radiation from a synchrotron and an electron storage ring, the basic principles and characteristics of insertion devices, namely wigglers and undulators, are described, including the delivery of both linearly and circularly polarised radiation. Key properties of the sources, namely their electron and photon emittance, are described, together with some aspects of their control and implications. The extent to which synchrotron radiation can be regarded as coherent, and the contrast with conventional and free-electron lasers (FELs) is described. Finally, the properties and limitations of key optical components delivering monochromatic synchrotron radiation down a beamline to an experimental end-station are introduced.

2.1 What is synchrotron radiation?

Unsurprisingly, synchrotron radiation is (or at least was—the definition has broadened), the radiation emitted by an (electron) synchrotron, while a synchrotron is an accelerator for charged particles (generally electrons or protons), developed for particle physics experiments involving high-energy collisions between elementary particles. Some insight into the nature of a synchrotron can be gained from a very brief description of its predecessor, the cyclotron, illustrated in figure 2.1(a).

In a cyclotron the charged particles travel within a thin wide cylindrical cavity, split electrically into two separate 'dees', all contained in a vacuum chamber and mounted between the pole pieces of a large permanent magnet. The presence of the magnetic field causes the particles injected perpendicular to the magnetic field to travel in circular orbits, but by applying an oscillating electric field between the two

doi:10.1088/978-0-7503-3847-9ch2

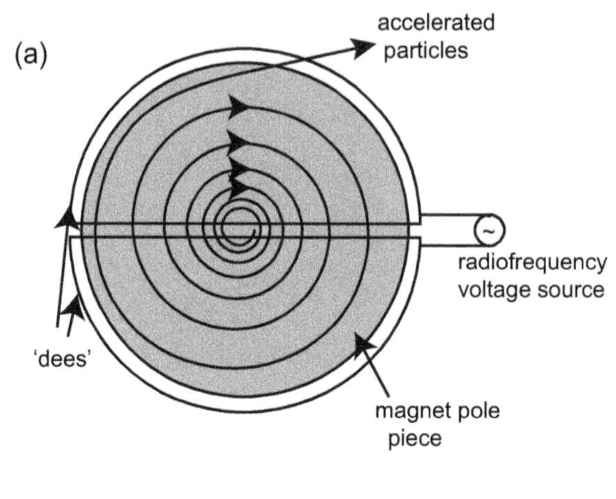

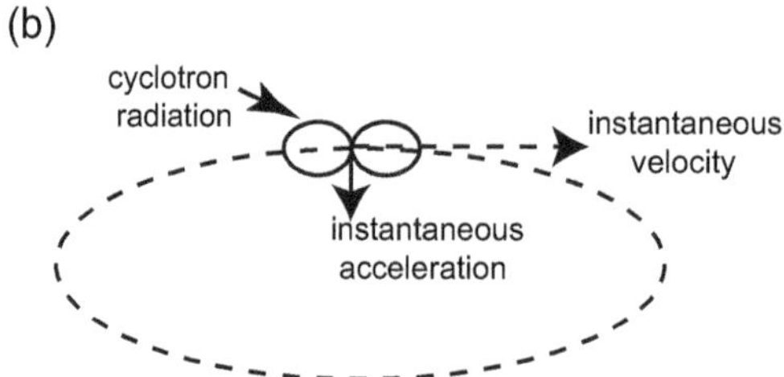

Figure 2.1. (a) Schematic plan view of a cyclotron, and (b) instantaneous power distribution of cyclotron radiation from an accelerated electron.

dees at a radio frequency (RF) matched to their circulating period, the particles see an accelerating electrostatic field each time they pass between the two dees. This causes the particles to spiral outwards as they gain energy until they are extracted from near the outer edge of the dees to be used in a particle physics collision experiment. One consequence of the circular motion of the charged particles is the emission of electromagnetic dipole radiation (as illustrated in figure 2.1(b)) with an angular dependence of the power distribution (for non-relativistic particle energies) proportional to $\cos^2\theta$, where θ is the angle of emission relative to the instantaneous direction of travel. The frequency of this emission is, at the lowest particle energies, that of the circulating particles. This RF 'cyclotron radiation' also occurs naturally beyond the ionosphere, emitted by high-energy electrons travelling in the Earth's magnetic field, but also in plasma in interstellar medium in locations where appropriate magnetic fields exist. Evidently, the maximum energy achievable in a cyclotron is limited by the size of the magnetic pole pieces and the strength of the magnetic field. In a series of increasingly large cyclotrons, built by Ernest Lawrence in Berkeley in the USA, one built in 1939 had pole pieces with a diameter of 94 cm and accelerated protons to 16 MeV. However, the cyclotron is not well-suited to electron acceleration because of their low rest-mass energy, so their total energy quickly becomes relativistic, causing their transit time between the dees to cease to be constant; a constant frequency of the applied voltages therefore ceases to continue to accelerate them as they pass between the dees. Recall that the rest-mass energy ($E = mc^2$) of an electron is only 0.511 MeV, some 2000 times smaller than that of a proton, so this relativistic effect arises at a much lower total energy for an electron than for a proton.

In order to achieve much higher particle energies and to overcome the problem created when the particle energy becomes relativistic, the synchrotron was developed. In this device a nominally circular energy-independent particle orbit is maintained (strictly speaking, it is a polygon with rounded corners) using a sequence of physically smaller high-field 'bending' magnets, separated by straight sections of the vacuum vessel in which the particles circulate. Acceleration is achieved in a RF cavity, installed at one point in the ring, in which the phase of the standing wave is such as to ensure acceleration of each 'bunch' of particles on each pass. In order to keep the orbit constant (rather than a spiral as in the cyclotron), the magnetic fields of the bending magnets are increased as the particle energy increases. A highly simplified schematic diagram of an electron synchrotron is shown in figure 2.2(a). The key components are the dipole bending magnets, the RF cavity, and a 'kicker'

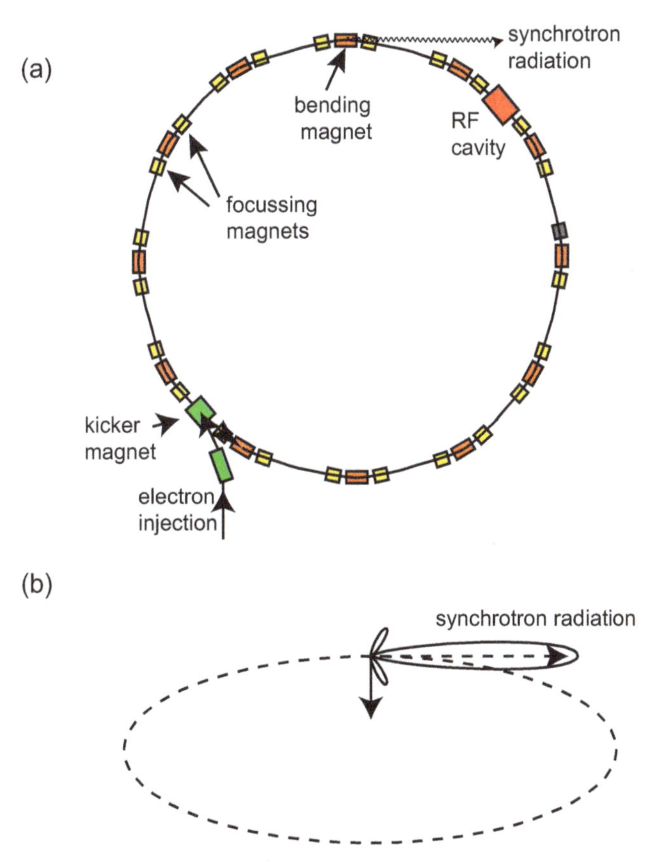

Figure 2.2. Panel (a) shows a simplified schematic of a synchrotron or an electron storage ring. Panel (b) shows the angular distribution of synchrotron radiation arising from a circulating electron at relativistic energies.

magnet to allow the initial electron pulses to be inserted into the nominally circular ring orbit. In addition are groups of mainly quadrupole focussing magnets arranged to achieve 'strong' or 'alternating gradient' focussing; the impact of this focussing of the electron beam on the characteristics of the emitted synchrotron radiation will be discussed in section 2.2. In a synchrotron, the typical accelerated particle energies are in the GeV range, so electrons are travelling at speeds very close to c and are highly relativistic. This has a profound influence on the nature of the emitted electromagnetic radiation. Firstly, a Lorentz transformation hugely distorts the angular distribution from the original standard dipole radiation, leading to a narrow cone of emitted radiation in the forward direction of the electron's travel through the dipole magnets (see figures 2.1(b) and 2.2(b)).

In discussing the properties of particles moving at relativistic energies it is generally convenient to express their speed, v, relative to the speed of light, by the parameter $\beta = v/c$ and to define an important parameter, γ, which is the ratio of the particle's total energy, E, to its rest-mass (m_0) energy:

$$\gamma = 1/\sqrt{(1 - \beta^2)} = E/m_0c^2.$$

The instantaneous angular width of the emitted radiation depends on the photon energy but is of the order of $1/\gamma$. Notice that as the rest-mass energy of an electron is approximately 0.5 MeV, γ for an electron is approximately $2000E$, when E is expressed in GeV. For a synchrotron source operating at 3 GeV this angular width of the emitted cone of synchrotron radiation is thus ~0.17 mrad or 0.01°. This radiation is emitted tangentially at each bending magnet in the ring; this is shown at just one of the bending magnets in figure 2.2(a).

The relativistic particle energy also leads to a huge 'blue shift' in the frequency of the emitted radiation as the emitting electrons approach the 'observer' in the instantaneous direction of travel, combined with excitation of multiple harmonics (with harmonic numbers as high as $\sim 10^{10}$). This leads to the broad high-photon-energy spectrum rather than the unshifted single RF emission expected at much lower particle energies; small variations in the energies of different electrons ensure this harmonic spectrum is smeared into a true continuum.

A formal analysis of the emission of synchrotron radiation, originally presented by Schwinger (1949), shows that if the angular frequency of the circulating particles is ω_0, the power radiated by a single particle at an angular frequency ω is

$$P(\omega) = \frac{3\sqrt{3}}{16\pi^2\varepsilon_0}\frac{e^2}{R}\gamma^4\frac{\omega_0}{\omega_c}G_1(\omega/\omega_c),$$

where R is the bending radius of the particle trajectory within the bending magnets and $G_1(\omega/\omega_c)$ is known as the universal curve of synchrotron radiation from bending magnets:

$$G_1(\omega/\omega_c) = \left(\frac{\omega}{\omega_c}\right)\int_{\omega/\omega}^{\infty} K_{5/3}(y)\mathrm{d}y.$$

$K_{5/3}(y)$ is a modified Bessel function and ω_c is known as the critical frequency, leading to a critical photon energy ($\hbar\omega_c = h\nu_c$). In practical units the bending radius can be related simply to the magnetic field, B, in the bending magnets by R (in metres) = 3.3E (in GeV)/B (in T), while

$$\hbar\omega_c = \hbar\frac{3}{2}\gamma^3\omega_0 = \hbar\frac{3c}{2R}\gamma^3.$$

Notice that the emitted power scales as γ^4 and thus as $(E/m_0)^4$. Evidently, this means that the maximum radiated power is achieved with the lowest particle mass. This is the reason for using electrons (or positrons) in a synchrotron radiation source rather than protons, which would radiate a factor of $\sim 10^{13}$ less power at the same value of γ. It is also clear that the total radiated power increases steeply as the electron energy increases, although other criteria determine the optimum energy for a synchrotron radiation source.

Figure 2.3 shows the universal curve for the synchrotron radiation flux emitted from a bending magnet in terms of the number of photons per second, normalised to the horizontal divergence in milliradians, the electron current in milliamps, the electron energy in GeV and per 0.1% spectral bandwidth. The energy and wavelength scales are normalised to the critical values of the device determined by the electron energy and the magnet bending radius, as shown in the equation. For example, at the 3 GeV Diamond facility in the UK, the bending magnet field is 1.4 T, which leads to a bending radius of 7.07 m and a critical energy of ~8.4 keV. One interesting property of the universal curve is that the total energy of the photon flux

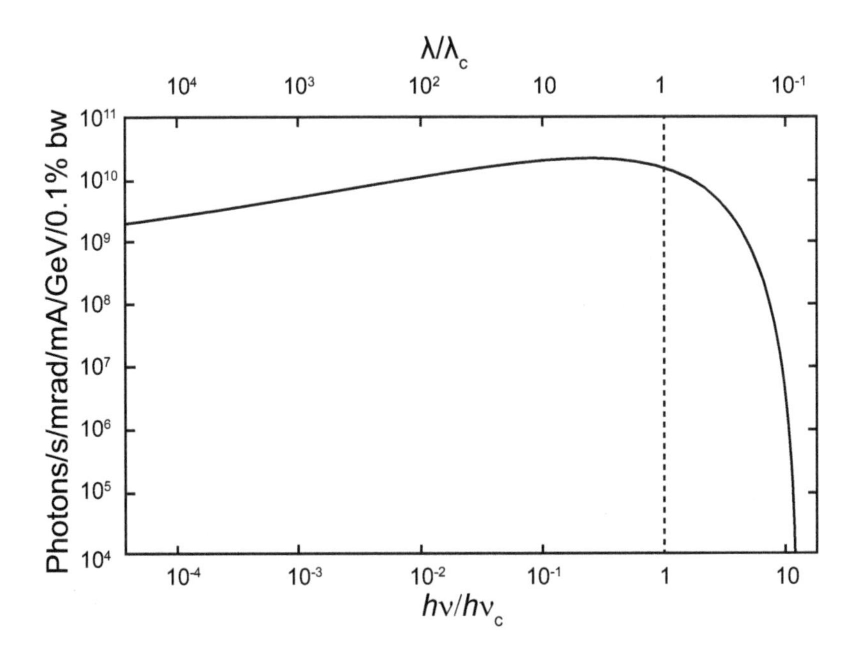

Figure 2.3. 'Universal' curve for synchrotron radiation from bending magnets based on normalised energy and emitted flux scales.

emitted below the critical energy equals that of the total flux of photons emitted above the critical energy.

Of course, synchrotrons were designed to accelerate particles to a high energy and then extract them to hit a target, in order to study the high-energy particle physics that results from these collisions. They operate in a cycle of injection, acceleration and extraction, typically performed at a frequency of a few tens of hertz, so the synchrotron radiation produced from such a machine is pulsed and its spectrum is time dependent as the electron energy is ramped up in each cycle. The earliest users of synchrotron radiation did work parasitically on these machines, while they were being operated in this mode for experiments in high-energy physics, but modern purpose-built synchrotron radiation facilities do not use synchrotrons but instead use high-energy electron storage rings. At the level of the highly simplified schematic diagram of figure 2.2(a), an electron storage ring and a synchrotron are identical, although there are certainly very important detailed differences in their electron optics relating to their primary use. In early electron storage rings built for use as sources of synchrotron radiation the ring was first used as a synchrotron, to accelerate electrons from a lower energy of injection to the final energy of the stored beam; thereafter, they operated as a storage ring, the RF power serving to re-accelerate those electrons that had lost energy in each circuit of the ring (most notably due to the emitted synchrotron radiation) to achieve a stable stored beam at constant energy. The electron current in such a stored beam does slowly decay, due to electron scattering from residual gas atoms and molecules, due to electron–electron scattering, and due to failures of the electron optics to refocus those electrons that stray too far from the planned mean trajectory around the ring. The half-life for this decay is generally of the order of a few hours, with both shorter and longer values being characteristic of the design parameters of each machine. Operating in this way, the synchrotron radiation flux slowly decays with time until a new injection is initiated. Nowadays, however, most synchrotron radiation storage rings operate with the electrons being injected at the full operating energy of the machine from a separate accelerator. This makes it possible to operate these sources in a 'top-up' mode, with small additional numbers of electrons being injected at regular intervals (of order minutes) to maintain a near-constant beam current and thus near-constant synchrotron radiation flux.

Although for most users synchrotron radiation can generally be regarded as quasi-continuous in time, it is in fact pulsed. This is an intrinsic characteristic of the way that the circulating electron energy is kept constant by the energy 'kicks' that each circulating electron experiences as it passes through the RF cavity. An electron that has lost a lot of energy through synchrotron radiation in its last transit around the ring needs to arrive at the RF cavity such that the phase of the RF standing wave in the cavity gives it a larger acceleration, whereas if it arrives having lost little energy it should arrive at a time that ensures it gains less energy from the RF wave. Evidently, this phasing requirement means that the circulating electrons cannot be distributed evenly around the ring, but instead travel in 'bunches'. This leads to periodic pulses of synchrotron radiation as each bunch passes through a bending magnet, the duration of the pulse being determined by the physical length of the electron bunch, while the periodicity of the pulses is determined by the RF frequency. The (maximum) number of electron bunches that circulate around the ring is determined by the time it takes an electron to travel around the ring (effectively at the speed of light) divided by the inverse of the RF frequency. This number depends on the circumference of the ring and the RF frequency it operates at, but is typically in the hundreds. In general, the high pulse rate and the relative long pulse length means that the time structure of the radiation is not suitable for most time-resolved scientific applications. For example, a typical RF frequency of ~500 MHz leads to a pulse separation of ~2 ns, but the pulse length is often ~tens of picoseconds. It is possible to operate a storage ring with fewer bunches (i.e., with some empty 'buckets' that could contain an electron bunch), to increase the time gap between radiation pulses—and indeed a range of different modes of operation of the storage ring can be used with groups of buckets occupied and others empty to allow larger time intervals between pulses of radiation to be used for time-resolved applications—but the ultimate example of this, when only a single bucket is filled ('single bunch operation'), is rarely exploited nowadays because of the huge decrease of average radiated power suffered by users not trying to perform time-resolved experiments.

2.2 Photon emittance, electron emittance and polarisation

The radiation emitted when electrons pass through the bending magnets of a synchrotron is the original definition of 'synchrotron radiation'. Nowadays, this is generally referred to as 'bending magnet radiation' to distinguish it from the radiation produced by the passage of the electrons through 'insertion devices'—namely wigglers and undulators, comprising groups of additional dipole magnets inserted into the straight sections between adjacent bending magnets, described below in section 2.3. Radiation from undulators has significantly different characteristics from bending magnet radiation but is now generally also considered to be 'synchrotron radiation'. This looser definition includes any radiation emitted by a modern synchrotron radiation facility source, comprising a synchrotron-like electron storage ring with a range of installed insertion devices. While use of these facilities is increasingly dominated by the radiation produced by insertion devices, an understanding of aspects of the characteristics of bending magnet radiation is relevant to the output from all these devices.

In the previous section, the influence of the relativistic energy of the electrons circulating a synchrotron or an electron storage ring on the angular distribution of the emitted radiation was highlighted. Specifically, the instantaneous emission of an electron travelling through a bending magnet was stated to have an angular width of $\sim 1/\gamma$, corresponding to a typical value of less than 1 mrad, although the true value depends on the photon energy (see below). Of course, each bending magnet causes the electrons to travel though a total angle of $360°/N_b$, where N_b is the number of (equivalent) bending magnets in the ring, typically about 20, so the total fan of emitted radiation in the horizontal plane (the plane of the electron orbit) is very much larger. In practice, the useable horizontal angular range is significantly less than this implied value of 15° (262 mrad) for $N_b = 24$, due to other components of the ring obscuring some of this range, and indeed the difficulty of designing optics that could transfer such a wide angle of radiation to an experiment down a beamline. Nevertheless, it is clear that bending magnet radiation comprises a radiation fan with a very small vertical divergence. The horizontal divergence emitted at a specific position in the bending magnets is also limited in a similar way, but if the optics transferring the radiation to an experiment can 'see' the emission from a range of positions as the electrons pass through the bending magnets the effective horizontal divergence may be much larger. As mentioned in the previous section, the intrinsic divergence actually depends on the emitted photon energy. Specifically, if the intensity spread within the total radiation fan is approximated by a Gaussian profile with a standard deviation σ_θ, the value of this parameter for a photon energy of $h\nu$ is

$$\sigma_\theta = \frac{0.565}{\gamma}(h\nu / h\nu_c)^{-0.425}.$$

Thus, at the critical energy the full width at half maximum, $2.35\sigma_\theta$, has a value of $1.33/\gamma$, while taking the width to be $\pm 2\sigma_\theta$ (with ~95% of the intensity within this range), the corresponding *half-width* is $1.13/\gamma$. However, at much lower photon energies (from the same storage ring with the same critical energy) this divergence is significantly larger. The fact that synchrotron radiation is intrinsically highly collimated clearly offers a huge advantage over a conventional laboratory X-ray source in terms of its ability to deliver a large number of photons to a sample. A standard laboratory X-ray source, based on the impact of high-energy electrons onto a solid target, emits X-rays over 2π steradians. Of course, the fact that the electron storage ring must be contained in a thick radiation-shielding wall generally means that it is rarely possible to get a sample much closer than about 10 m to the bending magnet source point, so if one wishes to study a small sample, or to investigate different microscopic parts of a sample, some form of focussing optics is required. The intrinsic limitations of what these optics can achieve is determined by the photon *emittance* of the source, defined as the product of the lateral size, σ_{ph}, and the lateral divergence, σ'_{ph}, of the source, because it is this quantity that is conserved in any optical design. This means, for example, that one can only decrease the focussed spot size of an optical source at the sample by increasing the convergence/divergence. This requirement to maintain a constant emittance (also referred to in light optics as the *etendue*) is known as the Helmholtz–Lagrange law and also applies to the electron optics within the electron storage ring.

Diffraction places an intrinsic limitation on how small the photon emittance can be. The classical single-slit diffraction pattern in light optics illustrates this clearly. The angular width of the central diffraction maximum is determined by the width of the slit and the wavelength of the light: a narrower slit (i.e., a smaller 'source') leads to a larger divergence of the transmitted radiation. Specifically, the diffraction-limited emittance for a source with a Gaussian distribution is $\sigma'_{ph}\sigma_{ph} = \lambda/4\pi$, where λ is the wavelength. In the case of synchrotron radiation, however, it is important to recognise that the emitted photons derive not from single electrons travelling along an ideal central path in the storage ring, but from bunches of electrons with a spread of slightly different trajectories. As each electron travels around the ring, it will lose variable amounts of energy, leading its trajectory to vary, the electrons with different energies being bent through different angles as they pass through the bending magnets. While passage through the RF cavity can re-accelerate these electrons, it is clear that at any specific location around the ring each bunch contains electrons with a (narrow) range of energies and trajectories. To minimise the impact of this problem and ensure that electron bunches continue to circulate around the ring for many hours, electron focussing lenses are installed in the ring. Specifically, a synchrotron uses 'strong' or 'alternating gradient' focussing, mainly based on quadrupole lenses but also including sextupole lenses.

As shown in figure 2.4, a quadrupole magnetic lens leads to focussing in one direction (the horizontal plane in this figure) and defocussing in the other (vertical) direction. Rotating this structure about the beam axis by 90° leads to the opposite effect. In the storage ring these two different device orientations are alternated around the ring. Somewhat counterintuitively, this combination of focussing and defocussing in both the horizontal and vertical directions leads to net focussing in both planes (the same effect is seen in light optics using alternating convex and concave lens with focal lengths of the same magnitude). While fuller details of the different types of focussing used in an electron storage ring are beyond the scope of

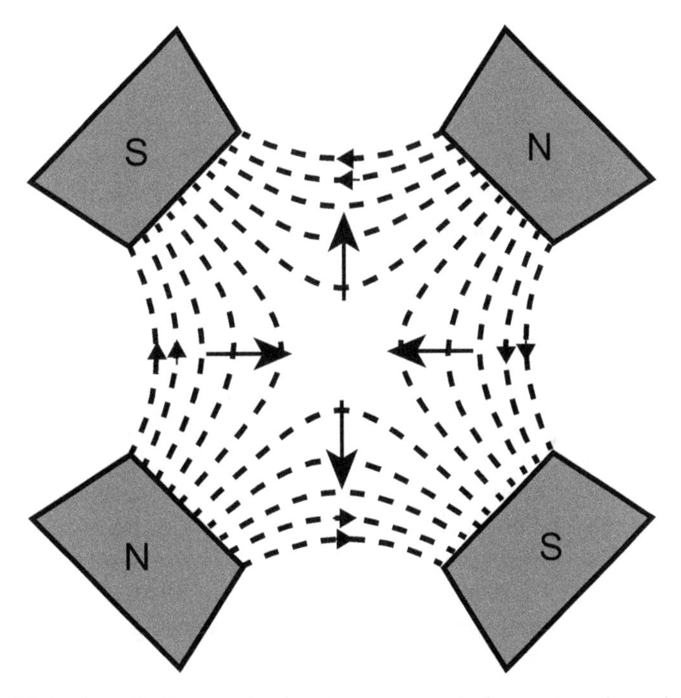

Figure 2.4. Simplified schematic diagram showing the arrangement of magnets and associated fields (small arrows) in a quadrupole focussing magnet. The large arrows show the directions of the forces acting on electrons passing through this device, leading to focussing in the horizontal plane of the diagram and defocussing in the vertical plane.

this book, there are two important implications for synchrotron radiation users, namely (i) that electron focussing optics are installed to minimise the *electron* emittance, and (ii) that this alternating gradient focussing means that while the electron emittance is constant around the ring, the two individual components, the lateral size σ_e and the lateral divergence σ'_e (in both vertical and horizontal planes), do vary significantly at different locations around the ring. This latter factor is important in identifying the best locations to install certain types of insertion devices. Of course, of even more direct importance to the synchrotron radiation user is that the emittance of the radiation to the user is determined by a combination of the electron emittance and the intrinsic, diffraction-limited photon emittance. The main thrust in the design of modern synchrotron radiation electron storage rings is to reduce the size of the electron emittance, commonly quoted as the product of the horizontal and vertical emittances, $\sigma_{eh}\sigma'_{eh}\sigma_{ev}\sigma'_{ev}$, which corresponds to the product of the area of the electron beam and its solid angle. A particular objective is to achieve a machine design that has an electron emittance equal to the diffraction limit of the photon emittance. Such a source is referred to as 'diffraction limited'. Of course, this diffraction limit depends on the photon energy at which the diffraction limit is calculated. The widely discussed objective is to achieve an electron emittance equal to the diffraction limit at a wavelength of the emitted radiation of ~1 Å (a photon energy of ~12 keV). No source has yet been designed to achieve this goal, although, of course, the diffraction limit is achieved in many sources at a sufficiently long wavelength. The driving force for such machine developments are thus experiments in the 'hard' X-ray range (photon energies above a few keV). As a measure of the success of these improvements, one typically quotes not the integrated photon flux output of a source (the quantity plotted in figure 2.3), but the spectral brightness (also often referred to as the brilliance), defined as the photon flux divided by the total emittance, i.e., the photon flux per unit area of the source per unit solid angle. This quantity is many orders of magnitude higher than the equivalent quantity from a conventional laboratory source and does reflect the potential to deliver very large numbers of photons to a small sample with minimal convergence or divergence.

Returning briefly to the case of cyclotron radiation emitted from an electron travelling in a circular path in the plane perpendicular to the direction of the imposed magnetic field, it is clear in this case that the emitted dipole radiation must be linearly polarised in this plane. The same is true for bending magnet radiation emitted from a synchrotron or electron storage ring in the plane of the electron motion. Within the plane of (horizontal) circulation of the electrons a detector 'sees' only motion within this plane, leading to emission of horizontally polarised radiation. Viewed from above or below this plane, however, the motion of the electron has a component perpendicular to the viewing plane, so there is a component of vertical polarisation, the two components summing to produce elliptically polarised radiation. Figure 2.5 illustrates this effect for the specific case of emission at a low photon energy of only 4.7% of the critical energy (150 eV). The effect occurs at all energies but the values of the out-of-plane angle, ψ, scale with the ratio of the photon energy to the critical energy and with the electron energy γ, so the

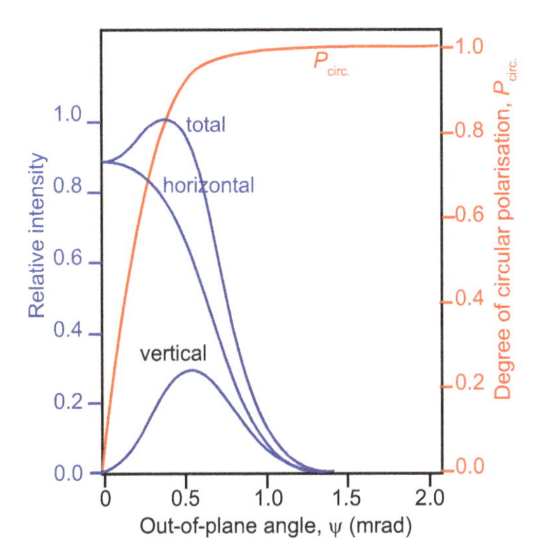

Figure 2.5. Relative intensities of horizontal and vertical polarisation of radiation emitted from a bending magnet as a function of the out-of-plane angle, ψ. Also shown is the resulting fraction of circular polarisation, $P_{\mathrm{circ.}}$ The photon energy for this plot is 4.7% of the critical energy. Reprinted from Finetti *et al* (2004), copyright (2004), with permission from Elsevier.

total angular range is much smaller at higher machine energies and photon energies, as described above. Notice, too, that the parity of the circularly polarised component switches above and below the radiating plane. Also shown in figure 2.5 is the fraction of circular polarisation, defined as

$$P_{circ} = 2\frac{A_h A_v}{A_h^2 + A_v^2},$$

where A_h and v_A are the corresponding amplitudes of the horizontal and vertical components. Notice that at large out-of-plane angles the polarisation is almost perfectly circularly polarised, but the intensity is vanishingly small!

2.3 Insertion devices

The broadband character of the radiation emitted from bending magnets shown in figure 2.3 clearly means that users at different beamlines, installed on identical bending magnets around a ring, have the possibility to select quite different photon energy ranges and work simultaneously on very different applications. However, in the most modern 'third-generation' synchrotron radiation facilities a large fraction of users (indeed, increasingly, the majority of users) now work on different beamlines at which the spectral output is more closely tailored to their specific requirements. This is achieved by inserting different magnetic structures in the straight sections of the ring between the bending magnets. The simplest such insertion device is a *wiggler*, also known as a wavelength shifter. The purpose of such a device is to produce a higher flux of radiation at the highest photon energies, and this is achieved by inserting a dipole magnet with a higher magnetic field than that of the bending magnets. This reduces the bending radius of the electrons that pass through this magnet, which increases the critical energy of emission, thereby shifting the universal curve of figure 2.3 to higher energies. As the emitted flux of this universal curve falls off quite steeply above the critical energy, shifting the curve to higher energy can lead to a large increase in the emitted flux at higher energies. Of course, as this magnet is installed in a straight section of the storage ring, it is essential that the electrons emerge from the inserted magnetic structure collinear with the trajectory they had when they entered the device. This can be achieved by installing lower-field steering magnets before and after the high-field magnet, as shown schematically in figure 2.6.

Even higher flux at higher energies can be achieved by installing not just one such high-field magnet in the straight section, but a multiple array of N such magnets, enhancing the emitted flux by a factor of N. In these *multipole wigglers* the high field may be achieved by using superconducting electromagnets.

The other general type of insertion device, the *undulator*, is superficially similar to a multipole wiggler in comprising a multiple array of dipole magnets, but operates at lower field strengths and has significantly different spectral emission characteristics. This type of device and the electron trajectory through it is shown schematically in figure 2.7.

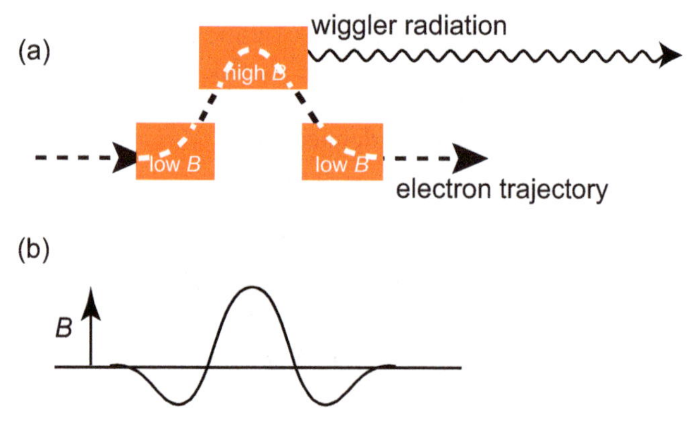

Figure 2.6. Simplified schematic diagram of a wiggler. Panel (a) shows the electron trajectory through the dipole magnets and the emitted radiation. Panel (b) shows the varying magnetic field experienced by the electrons along the electron trajectory.

The key difference between an undulator and a multipole wiggler is that whereas a multipole wiggler can be regarded as N separate sources of local high-field bending magnets, leading to a radiation intensity enhancement by a factor of N, in an undulator the amplitude of the radiation from these N magnets adds coherently, giving a potential intensity enhancement of N^2. The parameter distinguishing these two regimes is the 'undulator parameter', K:

$$K = \frac{B_0 e}{m_0 c} \frac{\lambda_0}{2\pi},$$

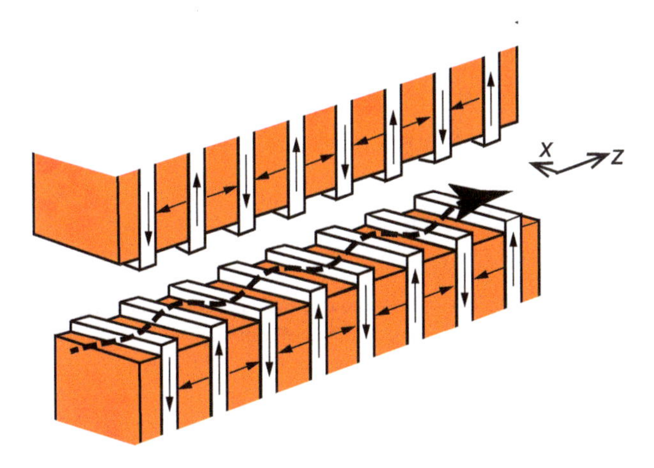

Figure 2.7. Schematic diagram of the magnet array of an undulator (or multipole wiggler), showing the electron trajectory through it (but with an exaggerated amplitude of the undulations from the axial trajectory).

where λ_0 is the physical periodicity of the magnet structure and B_0 is the amplitude of the magnetic field variation along the z-axis of the device, assuming that this is sinusoidal of the form $B = -B_0 \sin(2\pi z/\lambda_0)$. Notice that in practical units (B_0 in tesla and λ_0 in metres) $K = 93.36 B_0 \lambda_0$. The initial significance of this parameter is that it determines the amplitude of the horizontal angular deflection from the z-axis experienced by an electron passing through the magnetic array: specifically, $dx/dz = \frac{K}{\gamma}\cos(2\pi z/\lambda_0)$.

As the intrinsic opening angle of the emitted radiation from each magnet is $\sim 1/\gamma$, this means that if $K < \sim 1$, an 'observer' or detector viewing the emitted radiation along the undulator axis sees the emission from all parts of the electron trajectory. By contrast, if $K \gg 1$, the observer sees only the radiation from the electrons as they pass through the regions of maximum deflection and emit directly along the axis (see figure 2.8). These two extremes provide a distinction between the (approximate) conditions under which coherent summation of the emitted radiation amplitude from all parts of the separate magnets can, and cannot, add coherently and be emitted close to the axis of the device.

Figure 2.9 is a schematic diagram showing the condition for constructive interference of radiation emitted from equivalent positions in the periodic magnetic field of an undulator. Because of the lateral motion of the electrons as they travel through the field, the speed of the electron along the axis is reduced from its value in the absence of the periodic magnetic field, βc, to a value inside the device of $\beta_z c$. The transverse velocity of the electron is

$$\frac{dx}{dt} = \beta_x c = \beta c \frac{K}{\gamma} \cos\left(\frac{2\pi z}{\lambda_0}\right)$$

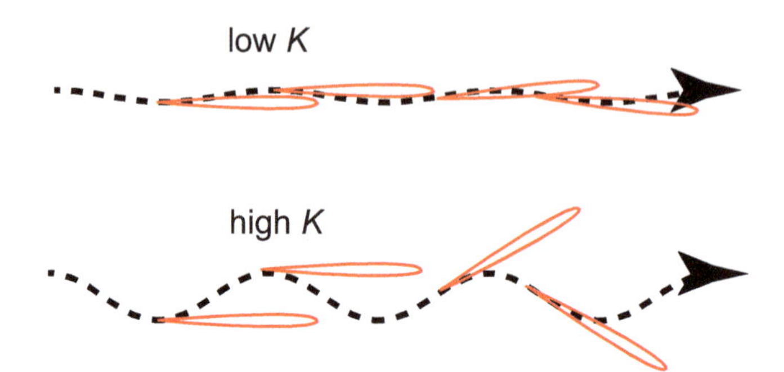

Figure 2.8. Schematic diagram showing instantaneous emission cones of radiation emitted from different parts of an electron trajectory in undulators of low- and high-K values.

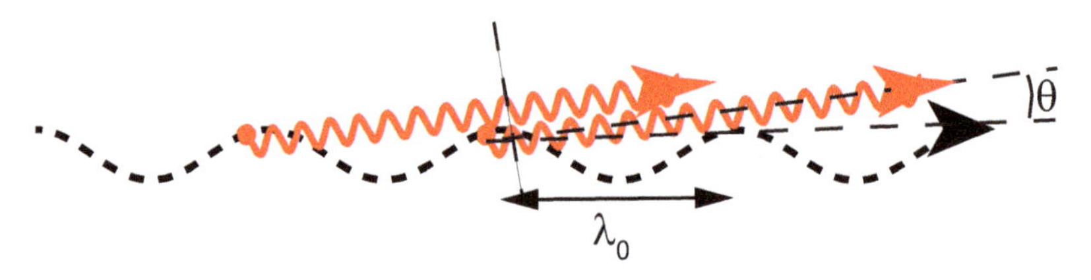

Figure 2.9. Schematic diagram showing the phase-matching condition of radiation from two equivalent points on the electron trajectory through an undulator.

and $\beta_z^2 = \beta^2 - \beta_x^2$, leading an expression for axial speed of

$$\beta_z^2 = \beta^2 - \frac{K^2}{\gamma^2}\left(\frac{1}{2} + \frac{1}{2}\cos\left(\frac{4\pi z}{\lambda_0}\right)\right).$$

Notice that the third term on the right shows that the electron velocity along the axis oscillates about its average speed of $\widehat{\beta_z}c$, given by the first two terms alone to yield

$$\widehat{\beta_z} \approx 1 - \frac{1}{2\gamma^2} - \frac{K^2}{4\gamma^2}.$$

Referring to figure 2.9, the separation of the wavefronts of the radiation of wavelength λ emitted at two neighbouring equivalent points on the axis is

$$d = \frac{\lambda_0}{\widehat{\beta_z}} - \lambda_0 \cos\theta.$$

And for constructive interference $d = n\lambda$, where n in an integer. This leads to the undulator equation defining the wavelengths at which this constructive interference occurs:

$$\lambda = \frac{\lambda_0}{2n\gamma^2}\left(1 + \frac{K^2}{2} + \theta^2\gamma^2\right).$$

Unlike the output from bending magnets or a multipole wiggler, an undulator therefore emits a spectrum dominated by intense harmonics defined by the parameter n, rather than the usual continuum of individual bending magnets. Notice that if K is very small the emission along the axis ($\theta = 0$) is at wavelengths equal to that of the periodic magnetic field but reduced by a factor $2n\gamma^2$. Recall that for a 3 GeV electron source $\gamma \approx 6000$, so this scaling factor is $\sim 10^8$, a factor that (approximately) transforms centimetres in the period of the undulator to ångströms in the emitted radiation wavelength. Notice one further feature of this undulator equation: increasing the value of K corresponds to increasing the strength of the periodic magnetic field, which increases the emitted wavelength and thus reduces the emitted photon energy. This is the opposite effect to that associated with increasing the magnetic field strength in a bending magnet, which shifts the critical energy (and thereby the whole emission spectrum) to higher values.

Examples of the on-axis output from a 5 cm-period undulator installed on a 3 GeV storage ring operating at several different low-K values are shown in figure 2.10, calculated using the SPECTRA computer program (Tanaka and Kitamura 2001). At the lowest value of K the output is dominated by a single peak at 1515 eV corresponding to the first harmonic ($n = 1$ of the undulator equation), although there is a very weak peak with 3 times this photon energy—the third harmonic—and an even weaker second harmonic peak. As K is increased, the first harmonic peak shifts down in photon energy, but higher harmonics (including even harmonics) gain very significantly in amplitude. The even harmonic peaks at low K are formally forbidden if the mean electron trajectory is exactly on the axis of the undulator and photon detection is also only on this axis. The weak peaks seen in the low-K spectra of figure 2.10 are due to the fact that the calculations assume a small divergence of the incident electron beam. The expected general absence of even harmonic emission on-axis can be understood by reference to figure 2.8. If K is sufficiently low that the angular deviations of the electron trajectory through the undulator are small compared with the opening angle of the instantaneous emission cone at each point of the trajectory, a detector placed on the axis of the undulators will see a continuous emission signal. The Fourier transform of a constant signal in time corresponds to a single discrete peak in frequency, so only the first harmonic of the undulator is observed. For a higher value of K, this axial detector only receives radiation when the electron passes through the positions of maximum positive and negative deflection, so the time dependence of the signal consists of a series of sharp peaks of alternating opposite amplitude. The Fourier transform of these sharp peaks now contains higher harmonics, but the fact that they are of opposite amplitudes means no even harmonics are involved. However, if the detector is placed off-axis, the alternating flashes of radiation are no longer equally spaced in time and even harmonics appear in the Fourier transform.

Evidently, considerable control of the spectral output of an undulator is available by adjusting the value of K, which in practice can be achieved readily by varying the gap between the magnet pole pieces. A large gap leads to a low value of B_0, and thus a low value of K; a small gap leads to a large B_0, and thus a large K. The maximum value of K achievable in this way is determined by the physical dimensions of the

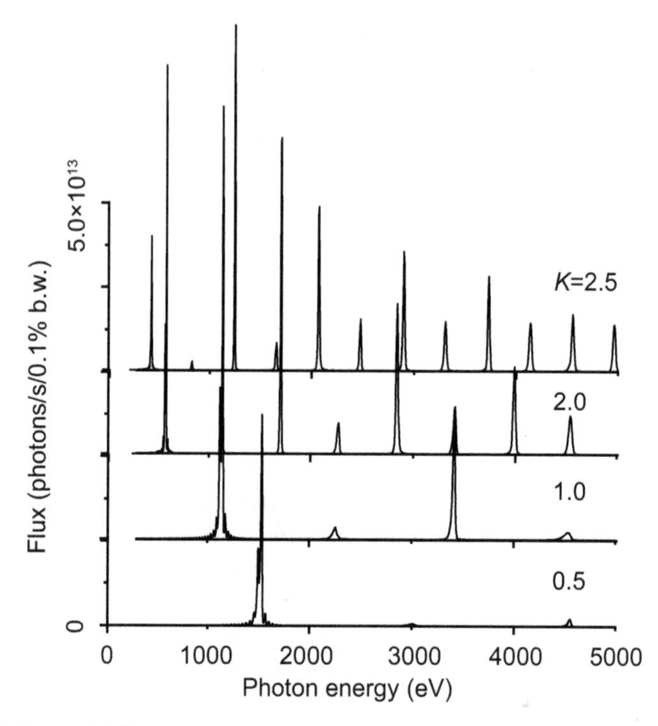

Figure 2.10. Calculated axial flux output through a small aperture at several different K values from a 2 m-long undulator with a 5 cm period installed in a 3 GeV storage ring with a circulating current of 300 mA.

vacuum vessel within the magnet or, for an in-vacuum undulator, the smallest gap that can avoid interfering with the stability of the circulating electron beam (of finite emittance).

The general utility of a particular undulator design installed on a specific storage ring can be determined by plotting the peak flux density of different harmonics over the full range of available K values. Calculated results in this form for the same undulator parameters as those used for figure 2.10 are shown in figure 2.11. This shows the output for the first five odd harmonics as a function of their photon energy. The K values associated with these settings are also shown. Notice that the sharp falloff in intensity at the highest energies corresponds to the lowest K values when the magnetic field strength B_0 becomes very low.

While the gain in flux density at the harmonics of an undulator, relative to bending magnet radiation, is considerable (and the gain in spectral brightness is even higher) the accessible range of photon energies from a single specific undulator is much narrower; in figure 2.11 the useful photon energy range covers a factor of ~× 30. Of course, undulators of different periods extend this range, but there are practical limits to these periods; at shorter periods it becomes difficult to achieve a periodic magnetic field of reasonable strength, while the need to close the gap between the pole pieces is constrained by the need to maintain a stable circulating electron beam. Moreover, much longer periods reduce the number of such periods that can be accommodated (and thus the degree of intensity enhancement) within a

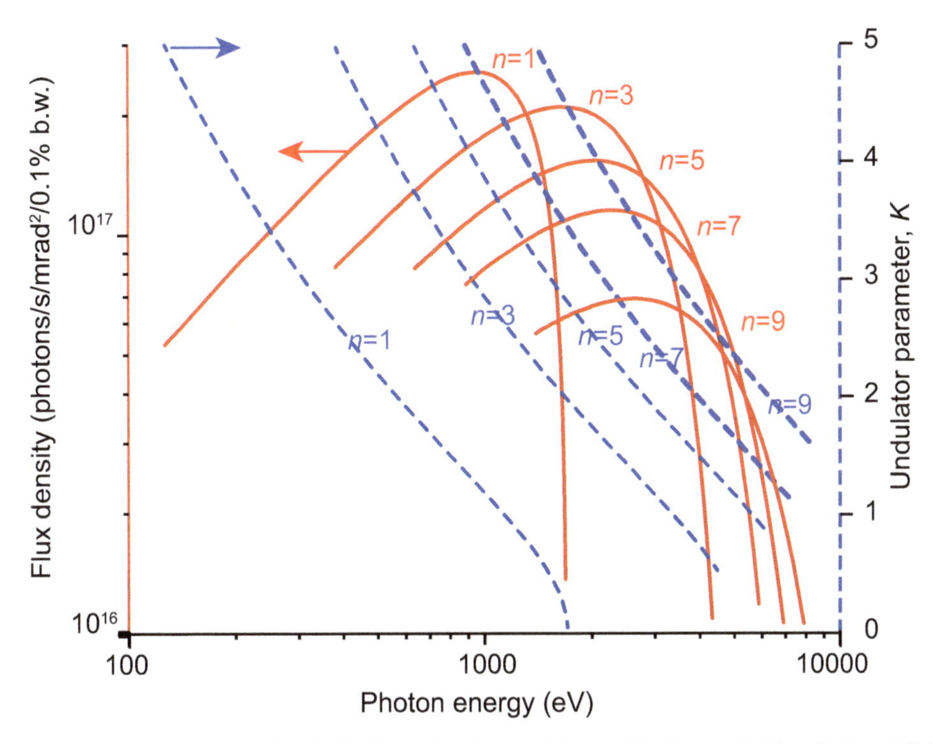

Figure 2.11. Calculated output flux density from a 2 m-long undulator with a 5 cm period installed in a 3 GeV storage ring with a beam current of 300 mA. For each harmonic the associated K values are also shown.

given total length of the straight section of the storage ring (typically no more than ~5 m). With the general upward trend in machine operating energies it becomes increasingly difficult to design undulators that are suitable for the delivery at the lowest photon energies down to 10 eV or less, relevant to some spectroscopic studies. In particular, as shown by figure 2.11, the lowest photon energies are obtained at the highest K values, but a high value of K leads to high intensity at higher photon energies (the device starts to become a wiggler), leading to high heat loads on the optics used to transfer the radiation to a sample.

Of course, while the broadband continuum of bending magnet radiation can service a very broad range of different users, in practice most surface science users (and indeed most users generally) want to use monochromatic radiation, albeit with the capability for tuning the wavelength over a wide range. Evidently, undulator radiation is energy-selected, so an important question is, how monochromatic are the different undulator harmonics? In particular, can they be used without the need for monochromators? The answer, in most practical situations, is no. To evaluate this, note that the criterion for constructive interference of the emission from two adjacent equivalent positions in the undulator was given above as $n\lambda = \frac{\lambda_0}{\bar{\beta}_z} - \lambda_0 \cos\theta$. Evidently, this is also the condition for constructive interference between equivalent positions at the two ends of the undulator and in the middle of the undulator (multiplying both sides of the equation by the number of periods N).

The condition for *destructive* interference between a position at one end of the undulator and the equivalent position at the middle of the undulator, leading to a zero amplitude, then occurs at λ', given by

$$\frac{N}{2}n\lambda' + \frac{\lambda'}{2} = \frac{N}{2}\left(\frac{\lambda_0}{\widehat{\beta}_z} - \lambda_0 \cos\theta\right),$$

leading to $Nn\lambda' + \lambda' = Nn\lambda$, so the wavelength range over which the harmonic emission occurs is $\lambda \pm \Delta\lambda$, where $\Delta\lambda/\lambda \approx 1/Nn$. For the 2 m, 5 cm-period undulator used for the results of calculation presented in figures 2.10 and 2.11, $N = 40$, so for the first harmonic $\Delta\lambda/\lambda \approx 1/40$, which is unlikely to suffice for any surface science spectroscopic studies. Notice, too, that this calculation relates to the axial emission alone. The wavelength emitted depends on the value of the emission angle θ, so in practice, when one accepts some off-axis radiation, the bandwidth of these harmonics may be significantly larger than this value.

A similar calculation evaluating the change in emission angle that leads to destructive interference and zero emitted intensity leads to a value of the opening angle of the emission of $\Delta\theta$ given by $\Delta\theta = \sqrt{2\lambda/N\lambda_0}$, so for the same example undulator described above ($N = 40$, $\lambda_0 = 5$ cm, $E = 3$ GeV) at the first harmonic $\Delta\theta$ proves to be ~60 μrad, very significantly narrower than the value for a bending magnet (at 3 GeV, $1/\gamma$ corresponds to an angle of ~170 μrad).

The polarisation of the radiation from this type of planar undulator shares with bending magnet radiation the property that the emission in the plane of the storage ring (and the plane of the undulations in the undulator) is linearly polarised in this horizontal plane. However, out of this plane the two types of source differ. Out-of-plane bending magnet radiation is elliptically polarised, but undulator radiation is linearly polarised in all directions, although the plane of polarisation varies out of the horizontal plane. A similar relatively complex pattern of out-of-plane polarisation directions is found for different harmonics, with the notable exception of even harmonics. For even harmonics, emission in directions displaced from the undulator axis in the vertical plane, but not the horizontal plane, has a vertical plane of polarisation. The origin of this effect may be found in the equations of the electron motion through the undulator presented above. The main focus of this discussion is the undulations of the beam in the horizontal plane (perpendicular to the magnetic field direction), which results in an oscillation of the speed of the electrons in the undulation plane, $\beta_x c$, but as the overall speed of the electron is unchanged this leads to a similar oscillation of the speed of the electron along the axis $\beta_z c$ relative to its average axial speed $\widehat{\beta}_z c$. It is this axial oscillation that gives rise to the vertical polarisation emission in the vertical plane. However, as the dipole emission resulting from these oscillations peaks in the vertical direction (before its relativistic shift), there is zero emission along the undulator axis, which corresponds to the direction of a node in the angular dependence of the emission.

As remarked above, users wishing to exploit circularly polarised radiation can use out-of-plane bending magnetic radiation (albeit 'contaminated' with some plane-polarised component), but a planar undulator emits only plane-polarised radiation in all directions. However, different undulator designs can provide pure-circularly polarised radiation. One way to achieve this is with a helical magnet structure. A conceptually simpler solution is to recognise that circularly polarised radiation can be obtained from summing equal amplitudes of vertical and horizontal polarisation with the appropriate phase relationship. This can be achieved by a combination of two planar undulators, one with an oscillatory vertical magnetic field (leading to horizontally polarised radiation as described above), the second with an oscillatory horizontal magnetic field (leading to vertically polarised radiation). A simple magnetic structure installed between these two linear undulators acts as a phase shifter to ensure the output is pure circular polarisation. Indeed, by modifying the phase shift introduced by this device, one can switch between left- and right-circularly polarised radiation. In practice, however, this solution is constrained by the fact that the horizontal electron emittance of synchrotron radiation storage rings is significantly larger than the vertical electron emittance, so it is difficult to use such small gaps in the horizontal magnetic field pole pieces.

A somewhat different approach that is now widely adopted is based on the Advanced Planar Polarised Light Emitter (APPLE) elliptically polarised undulator (or EPU) of Sasaki (1994), shown in figure 2.12. Superficially, this comprises a pair of conventional linear polarised undulators arranged side by side, but by relative movements the upper and lower sets of magnetic pole pieces along the axis, and of the front and back sets of magnets, one achieves complete control of the direction of the magnetic field variations along the axis of the device. This means that it is possible to obtain not only left- or right-circularly polarised radiation, but also linearly polarised radiation with any orientation of the polarisation direction, including pure horizontal and pure vertical.

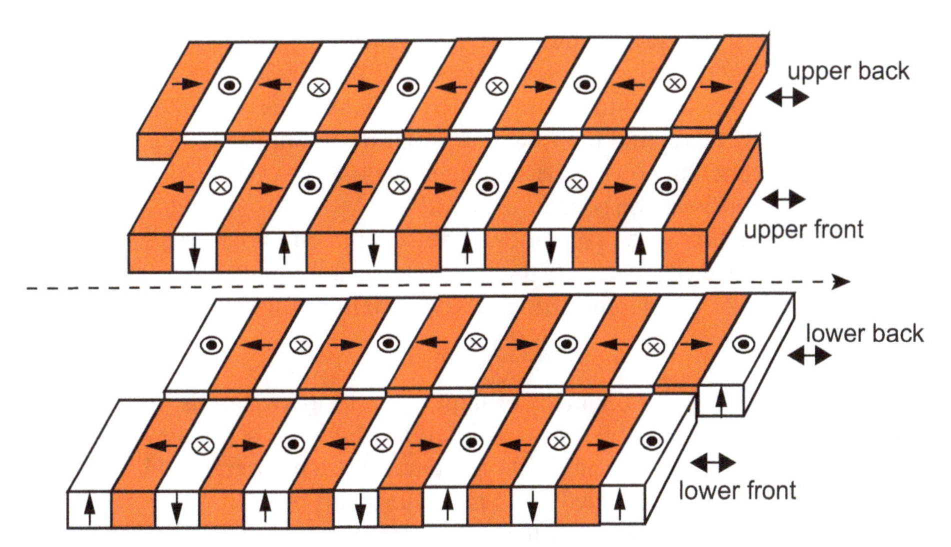

Figure 2.12. Schematic diagram of the arrangement of magnetic pole pieces in a section of the APPLE elliptically polarised undulator, together with the axial electron trajectory (not showing the superimposed trajectory undulations).

2.4 Coherence

In certain types of diffraction experiments an important property of synchrotron radiation is its coherence. In discussing this aspect of the radiation, it is important to recognise that synchrotron radiation involves the stochastic emission of individual photons by individual electrons with no correlation in time and phase between these events. This differs from the emission from a conventional laser in which the process of *stimulated* emission leads to photons that have a well-defined phase relationship to those of the stimulating field. Relative to this definition of coherence, synchrotron radiation is *not* coherent. Coincidentally, radiation from a FEL *is* coherent. FELs are sometimes regarded as 4th (or 5th?) general synchrotron radiation facilities in that their radiation is also generated by bunches of high-energy electrons passing through undulators. A key difference is that an FEL involves a linear accelerator through which the electrons pass only once, making it possible to generate much shorter electron bunches, but also to use very much longer undulators. In this case, the extremely intense emitted radiation does interact with the electrons to create 'micro-bunching' of the electrons, thereby generating truly coherent radiation. As in a conventional laser, this arises due to the interaction of the emitters with the electromagnetic field, but the mechanism is quite different. The effect in the FEL is related to the relative length of the bunches and the emitted wavelength, so this type of coherence can arise in a synchrotron radiation storage ring, but only at the longest wavelengths in the far-infrared.

Despite this, synchrotron radiation is increasingly referred to as being partially coherent, particularly as new machines are built with lower electron emittances, enabling a range of user experiments that rely on having a 'coherent source'. To understand this, it is helpful to recall the properties of the basic double-slit (Young's slits) experiment in visible-light optics.

This is shown in figure 2.13. In (a) a point source illuminates the two slits, and the properties of the diffraction pattern are determined by treating the two slits as coherent sources of Huygens secondary wavelets; the maxima and minima of the interference pattern correspond to the positions at which the optical paths from the two slits differ by integral and half-integral values of the wavelength, λ. In this case, the two slits are coherent sources because they are illuminated by different parts of the same wavefront from the source.

In figure 2.13(b), there are two point sources separated by a distance s leading to a displaced interference pattern such that the sum of intensities of the two patterns extinguish the visible interference fringes. The condition for this to occur is

$$\lambda/2d = s/D.$$

If one regards the two point sources of figure 2.13(b) as two parts of an extended source, this example highlights the key difference between a point source and an extended source in terms of their ability to demonstrate interference phenomena. If the size of the source (represented by s) is smaller than the value corresponding to this 'blurring condition', then it can be regarded as coherent in the sense that it allows coherent interference phenomena to be detected. In the foregoing text it was argued that individual photons emitted in synchrotron radiation have no well-

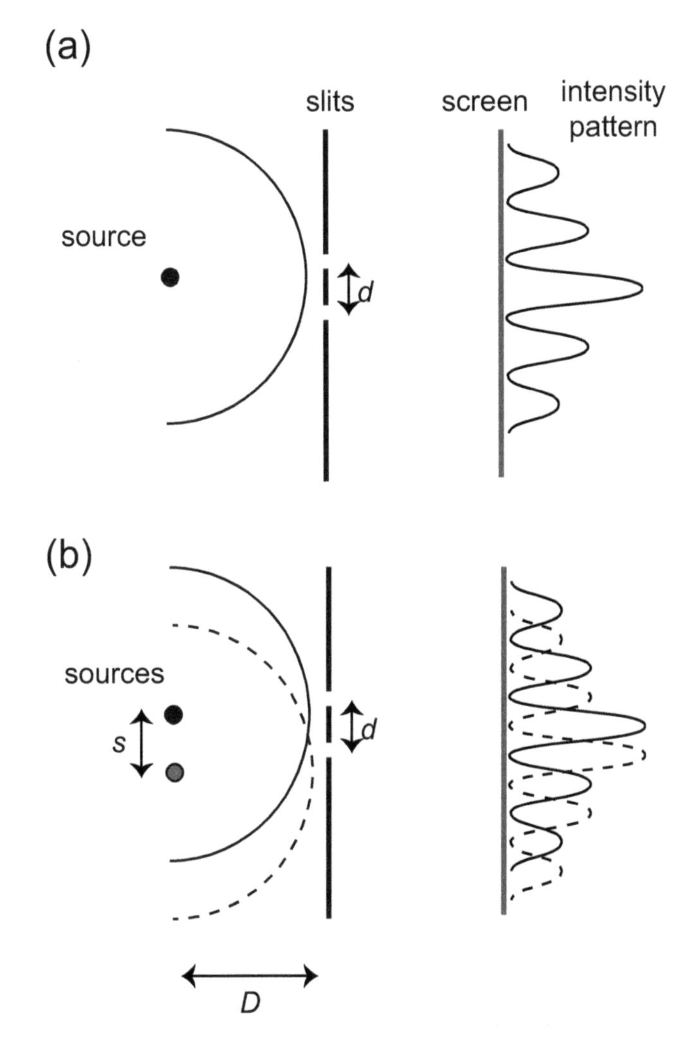

Figure 2.13. Simple schematic of the Young's slits experiment in optics. In (a) a single point source illuminates the two slits separated by d at a distance D, leading to the interference pattern shown schematically on the right. In (b) two point sources with a separation s (representing two positions in an extended source) lead to separate interference patterns displaced by one half of the periodicity of the component diffraction patterns, smearing out the observed pattern.

defined phase relationship, which is true, but if the physical dimensions of the source of these photons is small enough, interference effects are seen because separate components of the wavefield of each photon can interfere coherently. This condition leads to a definition of a '*transverse* coherence length', ξ_T, given by

$$\xi_T = \lambda D/2s.$$

As can be seen from figure 2.13(b), the offset of the component interference patterns that causes the visible effect to be obscured can arise either from an offset in position of the separate components of the source, or from an offset in the direction, so in effect the coherence length is determined by the degree of collimation of the source, which can be improved by inserting a small 'pinhole' aperture (with the evident loss of flux) for experiments that require a particularly large coherence length. As the source emittance is the product of the source size and its divergence, low-emittance sources have larger transverse coherence lengths. This leads to a definition of the 'coherent fraction', F, of a source, as the ratio of the ideal (diffraction-limited) emittance and the actual emittance:

$$F = \frac{(\lambda/4\pi)^2}{\Sigma_x \Sigma_{x'} \Sigma_y \Sigma_{y'}},$$

where the denominator is the product of the lateral sizes and divergence of the source in the horizontal and vertical directions. The *longitudinal* coherence length is determined by the monochromaticity ($\Delta\lambda$) of the source, which is determined by the monochromator used in the beamline rather than for the electron storage ring itself; this can be shown to be given by

$$\xi_L = \lambda^2/2\Delta\lambda.$$

2.5 Beamline optics

The mention of a monochromator here is a reminder that, while the design of the electron storage ring and its insertion devices determines the basic source characteristics, what is of interest to a synchrotron radiation user are the properties of the radiation when it arrives at their experiment. Although synchrotron radiation is intrinsically in the form of a narrow cone, the fact that an experiment cannot be placed much closer to the actual radiation source than 10 m or so, due to the tangential nature of the emission from the source that must be contained in thick (mainly concrete) radiation shield walls, means that some form of transfer optics, usually involving some degree of focussing, is essential if a small radiation 'spot' is to be achieved on a sample. In the visible-light wavelength range this would typically be achieved using refractive lenses, but most glasses absorb strongly in the ultraviolet and even fabricating optics from LiF leads to strong absorption for photon energies above ~10 eV. As the vast majority of applications of synchrotron radiation in surface science require much higher photon energies, from the vacuum ultraviolet (VUV) through soft X-rays (SXRs) and even into the hard X-ray range ($h\nu >$ ~10 keV), conventional refractive optics are therefore no use, and one must exploit reflective optics.

A key factor determining the constraints of reflective optics is that while in the visible wavelength range the refractive indices of lens materials are greater than unity (typically values for glasses are ~1.3–1.5), in this much higher energy range the refractive index, n, is slightly less than unity. This is usually written as

$$n = 1 - \delta + i\beta,$$

where δ is the deviation in the real part from unity and β determines the magnitude of the absorption expressed as the imaginary part of the refractive index. These terms can be related to the real and imaginary parts (suffices r and i) of the atomic scattering factors for forward (zero angle) scattering factors, f^0, of the constituent atoms of the material, as a function of the angular frequency, ω,

$$n(\omega) = 1 - \frac{n_a r_e \lambda^2}{2\pi}[f_r^0(\omega) - if_i^0(\omega)],$$

where n_a is the atomic density and r_e is the classical electron radius:

$$r_e = \frac{e^2}{4\pi\varepsilon_0 mc^2}.$$

An important consequence of the refractive index being less than unity is that the usual condition of total internal reflection experienced by light in the near-visible spectral range passing through a medium such as glass ($n > 1$) becomes the condition for total *external* reflection when $n < 1$. This condition is obtained from the usual Snell's law,

$$\sin\theta_{refr} = \sin\theta_i/n,$$

where the angles of incidence and refraction, θ_i and θ_{refr}, are measured relative to the interface normal. The critical incidence angle, θ_c, occurs when the refracted beam travels parallel to the surface, i.e., $\theta_{refr} = 90°$, so if there is no absorption ($\beta = 0$) then $\sin\theta_c = 1 - \delta$.

As this critical angle corresponds to grazing incidence, it is more helpful to recast this equation in terms of the grazing angle, $\phi_c = 90 - \theta$, whence

$$\cos\phi_c = 1 - \delta,$$

and as this grazing angle is small, taking the leading term of the series expansion of $\cos\phi_c$ leads to $\phi_c \simeq \sqrt{2\delta}$.

While this simple formula allows one to determine the critical angle, the implication that the reflectivity is unity ('total reflection) for grazing angles smaller than this value is only true if there is no absorption ($\beta = 0$). Although there is no transmitted (refracted) beam passing into the reflecting medium for smaller grazing angles, there is an evanescent wave that penetrates this medium, so absorption can occur. The true value of the reflectivity can be determined from the usual Fresnel equations, describing the incident, reflected and refracted waves and the matching conditions at the interface for the two perpendicular polarisation components, *p*-polarisation (with the polarisation vector in the plane of incidence) and *s*-polarisation (with the polarisation vector perpendicular to this incidence plane). Much the most widely used mirror coating for beamline optics is gold, due to its corrosion resistance and its high atomic number, which ensures not only strong scattering but also that there are large energy gaps between the absorption edges in the X-ray energy range. Figure 2.14 shows the calculated reflectivity of gold as a function of incident photon energy at a range of different grazing incidence angles, generated using an online tool[1]; the calculations are based on updated values of the atomic scattering factor first tabulated by Henke *et al* (1993).

[1] Available online at http://henke.lbl.gov/optical_constants/mirror2.html.

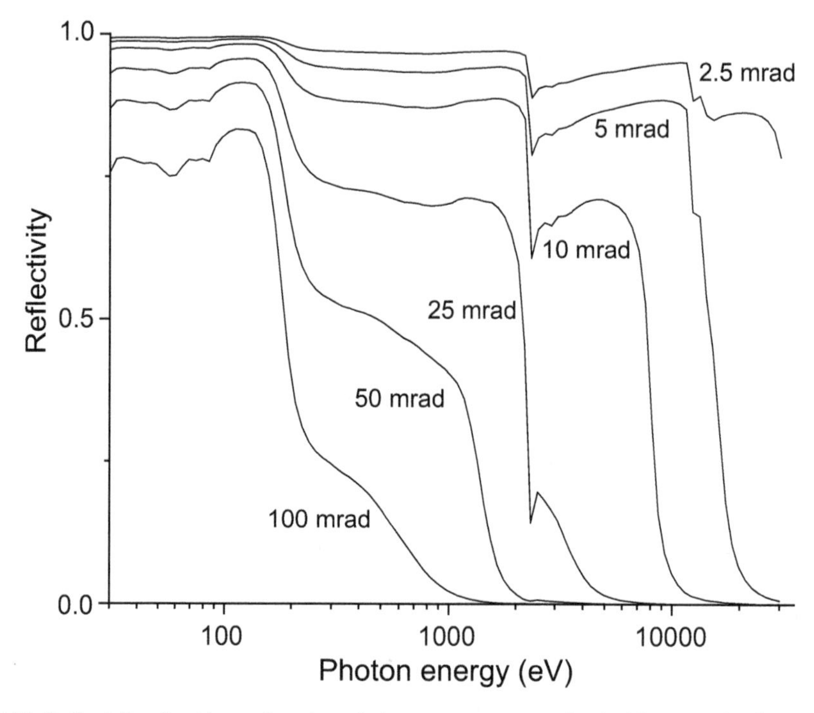

Figure 2.14. Reflectivity of gold as a function of photon energy at grazing incidence angles from 2.5 mrad (0.143°) to 100 mrad (5.7°). Based on data from https://henke.lbl.gov/optical_constants/mirror2.html.

As may be expected, particularly at the highest photon energies, the reflectivity is only high at very small grazing incidence angles, and even at an angle of 2.5 mrad (0.143°) there are clear drops in the reflectivity as the photon energy crosses the Au M_3 and M_2 absorption edges around 2200 eV (the thresholds for $3d_{5/2}$ and $3d_{3/2}$ photoemission) and the L_3 and L_2 absorption edges at ~12 000 eV and ~13 700 eV (thresholds for $2p_{3/2}$ and $2p_{1/2}$ photoemission). The impact of these absorption edges on the reflectivity increases significantly as the grazing incidence angle increases. Clearly Au is not a suitable mirror material for experiments requiring photon energy scanning close to these absorption edges, but there are wide energy ranges in which it performs well; if the ranges close to the Au absorption edges are important for specific experiments, an alternative coating (most commonly platinum), with displaced adsorption edges, can be used. More generally, figure 2.14 quantifies the range of grazing incidence angles that can be used for reflective optics in different photon energy ranges, without suffering large losses. However, this figure also demonstrates that mirrors act as 'low-pass filters'. For example, a grazing incidence angle of 1° (17 mrad) has quite high reflectivity for photons with energies less than ~2 keV, but reflects very few photons with significantly higher energies. This is a potentially valuable property for experiments requiring the lowest photon energies, for which undulators may have to operate at high values of K, leading to a large accompanying flux of higher-energy photons. A suitable choice of a pre-mirror reflection angle can strongly attenuate this undesirable flux of the high-energy photons, which would otherwise lead to a high heat load on subsequent optical components.

Of course, to achieve focussing the reflective surface must be curved. Analysis of reflection from a spherical (concave) surface, shown schematically in figure 2.15, identifies the main relevant phenomena. Specifically, while the focal length of a spherical mirror operating at normal incidence is *R*/2, where *R* is the radius of curvature of the mirror surface, at grazing incidence this focal length changes and is different in the tangential or meridional planes (the plane of incidence). These effective focal lengths are, in the tangential plane,

$$f_T = (R \sin \phi)/2,$$

and, in the perpendicular sagittal plane,

$$f_S = R/(2 \sin \phi).$$

For small angles of incidence these two focal lengths different enormously; for example, at 5 mrad incidence this ratio is 40 000. This effect is known as astigmatism. Under these conditions there is essentially no sagittal focussing, and the spherical mirror effectively acts as a cylindrical mirror. Notice, too that because the surface of the mirror is curved the angle of incidence differs at different parts of the incident beam 'footprint' on the mirror, leading to spherical aberration. The astigmatism can be overcome by using a mirror with a toroidal surface, like the

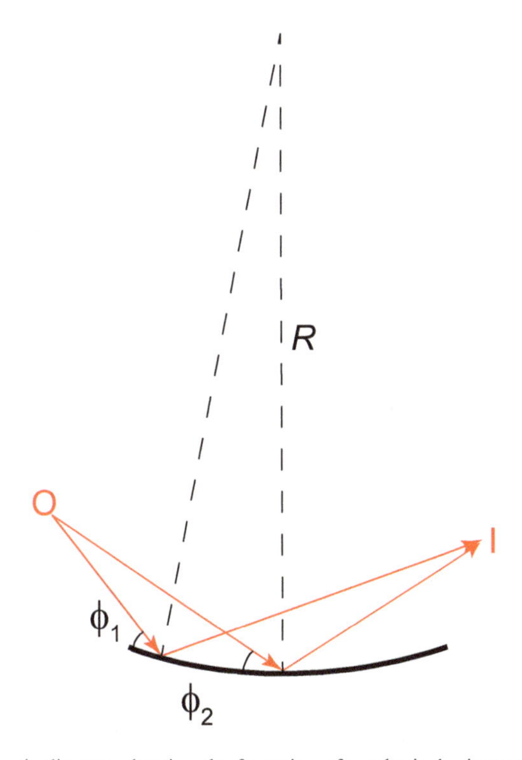

Figure 2.15. Schematic diagram showing the focussing of a spherical mirror at grazing incidence.

inside of a bicycle tyre, having very different radii of curvature in the two orthogonal planes, but the more commonly used solution is to focus separately in the meridional and sagittal planes using two cylindrical or spherical mirrors rotated by 90° about the optical axis, an arrangement known as a Kirkpatrick–Baez (or K–B) pair. There are extra reflectivity losses associated with the two reflections, but the manufacture of precise cylindrical and spherical mirrors is significantly simpler than that of toroidal mirrors.

While these grazing incidence mirrors are the most common beamline focussing optics, other methods can be used in some specific applications. In introducing reflective optics above, the possibility of refractive optics was discounted. Lenses made from different types of glass or crystalline compounds, such as LiF, function in the near-visible range of the spectrum, exploiting weak absorption and a refractive index greater than unity. By contrast, as described above, at higher photon energies many materials may be weakly absorbing but have an index of refraction less than unity. Of course, the focussing properties of refractive lenses rely not on their absolute value of the refractive index, but the difference in refractive index of the lens and the surrounding material (which is usually air or vacuum with a refractive index of unity). As a result, if the radiation is passing through a material with a refractive index of less than unity, refraction can occur when the radiation passes out of the material into air or vacuum. The effect is clearly weak, but an array of such 'vacuum lenses', referred to as a compound refractive lens, can provide useful focussing. The simplest way to achieve this effect is by drilling a series of holes in the transmissive material, producing cylindrical lenses with focussing in one plane, and is illustrated in figure 2.16. A single lens component, comprising a concave cylindrical lens of the transmissive material with a refractive index of $1-\delta$ and a radius of curvature of R, leads to a focal length of $R/2\delta$, while a series of N such components gives a refractive index of $R/2N\delta$; figure 2.16 shows an example of five such lenses. Of course, δ is very small, so a large number of these lens components (e.g., >100) is required for effective focussing, but a large number implies a long pathlength of the radiation through the transmissive material, potentially leading to

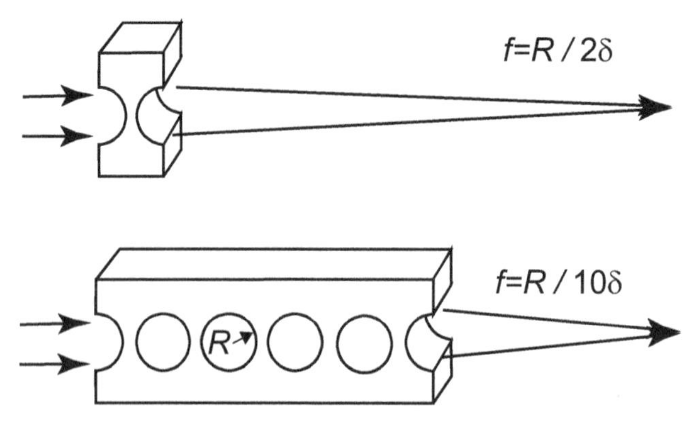

Figure 2.16. Schematic diagram illustrating the principle of X-ray focussing using compound refractive cylindrical lenses for the case of a single lens and a sequence of five lenses.

large absorption losses. This favours the use of low atomic number elements, such as beryllium, for the fabrication of these devices. Of course, unlike reflective optics, the focal lengths of refractive optics do depend on the photon energy (i.e., they suffer from chromatic aberration), and so are only well-suited to experiments working at a single photon energy.

A quite different type of focussing device, based on grazing incidence reflections, is provided by compound capillary optics. The underlying principle is the same as that of fibre optics used for infrared transmission lines, although for X-ray optics the individual components are hollow glass capillaries in which the X-rays are guided by grazing incidence reflections at the internal walls of the capillaries. Focussing is achieved by using tapered capillaries bent into arrays, as shown schematically in figure 2.17.

One further focussing device based on entirely different physical principles is the Fresnel zone plate (FZP), illustrated in figure 2.18, a device used at a number of angle-resolved photoelectron spectroscopy (or ARPES) beamlines to access very small regions of a surface, but also for spatial imaging by scanning the relative lateral positions of the sample and FZP. The FZP comprises a set of concentric rings of alternately nominally transparent and opaque material, the radii of these rings corresponding to locations that differ in their distance from a focal point by one wavelength of the radiation to be focussed. Radiation passing through the transparent rings arrives at the focal point in phase and interferes constructively, whereas radiation arriving at the opaque rings, which would be out of phase if it reached the focal point, is obstructed. The FZP is thus a special type of diffraction grating for near-field (Fresnel) diffraction that leads to focussing at a single point.

A simple calculation of the requirements for these in-phase and out-of-phase path lengths leads to an equation for the effective focal length as a function of diameter of the outermost ring, D, the width of the outermost zone, Δr, and the wavelength:

$$f_m \simeq D\Delta r/m\lambda,$$

where m is the diffraction order; as in a standard (Franhofer) diffraction grating comprising multiple linear parallel 'slits' (see below), the condition is satisfied by higher orders, with the wavelengths related by integral factors. These higher orders

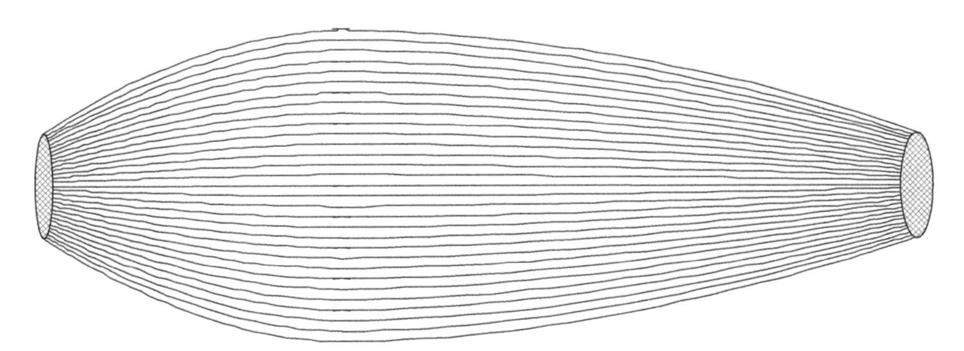

Figure 2.17. Schematic diagram of a monolithic polycapillary optic with a short input and longer output focal length. Reproduced from MacDonald (2011).

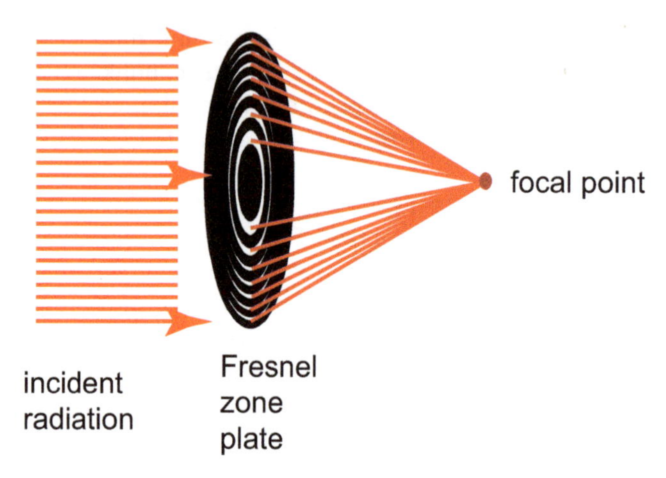

Figure 2.18. Schematic diagram showing the focussing of incident monochromatic radiation at a focal point by a Fresnel zone plate.

(and zero order) can be suppressed by an axial beam stop and order-sorting apertures. Of course, the device suffers from strong chromatic aberration and is thus only suitable for focussing monochromatic radiation (albeit with some incident higher-order radiation delivered by the monochromator, as described below). The theoretically (diffraction-limited) achievable focussed spot size is $1.22\Delta r$, limited in practice by the ability to fabricate a device with a sufficiently narrow outermost zone, typically achieved using lithography. In practice, this also limits the overall diameter of FZPs to sub-millimetre dimensions, which are then typically illuminated by radiation that is passed through a small aperture; this also ensures good coherence of the illumination across the FZP. A further constraint in the manufacture of these devices is that the 'opaque' rings do need to be strongly absorbing to achieve the necessary contrast, a requirement that favours their application using SXRs (and even VUV radiation) that are strongly absorbed in most materials. Under these conditions, the working distance (i.e the value of f_1) is only of the order of a few millimetres or less. Nevertheless, focussing down to a few tens of nanometres is possible.

2.6 Monochromators

The other key optical component in experimental beamlines, with beamline focussing located before or after (or both), is the device to extract monochromatic radiation from the spread of wavelengths of the incident synchrotron radiation (including undulator radiation). Very few experiments across the broad range of scientific applications explored at a synchrotron radiation facility are able to exploit the 'white' radiation.

Essentially, all monochromators operating in the broad energy range from the VUV (photon energies of a few tens of eV) to hard X-rays (mainly photon energies above ~8 keV) exploit diffraction in a reflection geometry from a periodic structure. For photon energies up to ~2 keV these are diffraction gratings comprising a one-dimensionally periodic set of 'lines', often mechanically 'ruled' (effectively scratched or machined) but also fabricated using lithographic techniques, whereas for higher photon energies the need for periodicity of the same order as the short wavelengths (mostly ~1 Å or less) means that the diffracting objects are crystalline solids. However, the fact that crystalline solids are three-dimensionally periodic (as opposed to one-dimensionally periodic diffraction gratings) does lead to some important differences in these two distinct types of monochromators.

Of course, monochromators or spectrographs operating in the visible or near-visible range of the spectrum can use transmission gratings, which comprise a periodic array of slits. Constructive interference of the light transmitted by successive slits occurs when their path lengths to a detector differ by an integral number of wavelengths, thus leading to radiation of different wavelengths being dispersed to different emergent angles. For normal incidence to a transmission grating with a periodicity d, one obtains the condition

$$d \sin \beta = m\lambda,$$

where β is the angle of transmission relative to the normal incidence direction (i.e., the complement of the grazing incidence angle ϕ) and m is the diffraction order; in general, only the first order ($m = 1$) is used. Notice that $m = 0$, zero order, corresponds to a transmission that is independent of wavelength, so 'white' light is transmitted with $\beta = 0$. In practice these 'slits' are simply the spaces between opaque lines on a transmissive substrate, which must be weakly absorbing to the radiation of interest. As such, this approach is not possible at photon energies higher than a few eV in the VUV and SXR range. A reflection grating operates in the same general fashion but requires only that the surface has alternating stripes of relatively low and high reflectivity. As discussed in the context of mirrors in section 2.5, high reflectivity at high photon energies requires grazing incidence, so while monochromators based on reflection gratings using near-normal incidence can be used for photon energies up to ~30–40 eV, at higher energies more grazing angles are necessary to achieve a reasonable degree of transmission through the monochromator. For incidence at an angle of α relative to the normal to the average grating surface, the diffraction condition is

$$d(\sin \alpha + \sin \beta) = m\lambda,$$

where the angle of outgoing diffracted radiation, β, is also defined relative to the normal to the average grating surface, as shown in figure 2.19. Notice that zero-order diffraction occurs when $\sin \alpha = -\sin \beta$, so the sign convention for β is that it is negative when the scattered beam is on the opposite side of the surface normal to the average grating surface from the incident beam (see the + and – signs in figure 2.19). Reflection gratings are commonly 'blazed', meaning that the surface is cut with a sawtooth profile, the diffractive/reflective surfaces being tilted relative to the average grating surface by a blaze angle, γ. While the condition for coherent interference between scattering (diffraction) from the individual blazed terraces is derived by assuming the interference of 'Huygen's secondary wavelets' from these regions, a true reflection from these tilted surfaces requires that the angle of reflection equals

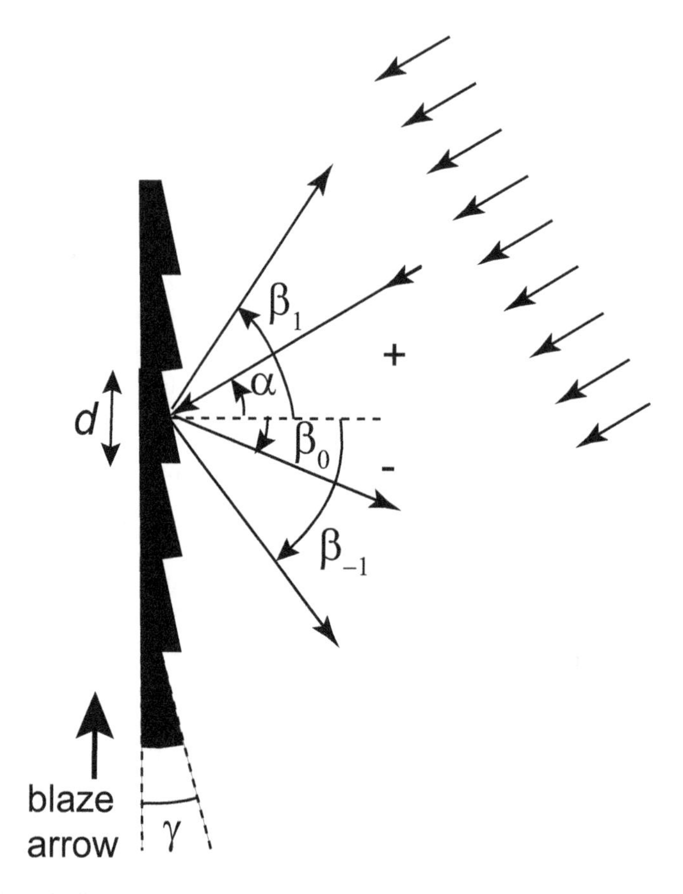

Figure 2.19. Schematic diagram defining the parameters associated with scattering from a blazed reflection grating with a periodicity d and a blaze angle γ.

the angle of incidence *relative to these blazed terraces*. Under this condition the overall scattered yield (the efficiency of the grating) is expected to be greatest. This corresponds to a particular wavelength of the diffraction equation. The blaze angle is thus chosen to optimise performance in the range around a particular wavelength. Other surface profiles, such as lamellar and sinusoidal, may be used for particular applications or fabrication methods.

Using a plane grating, as illustrated in figure 2.19, the incident parallel beam of radiation is dispersed in different directions depending on the wavelength, but to isolate one specific wavelength (strictly a narrow range of wavelengths) the dispersed output needs to be brought to a focus such that a suitably placed aperture selects the wavelength angle of interest. Figure 2.20 shows the minimum requirements for a grazing incidence plane grating monochromator (PGM). The source-to-monochromator distance is large (generally greater than 10 m) due to the shield wall and various 'front-end' beamline components to maintain the integrity of the ultra-high vacuum (UHV) of the storage ring, to obstruct stray radiation from various electron collisions that can occur in the ring, and movable beam stops to allow the radiation

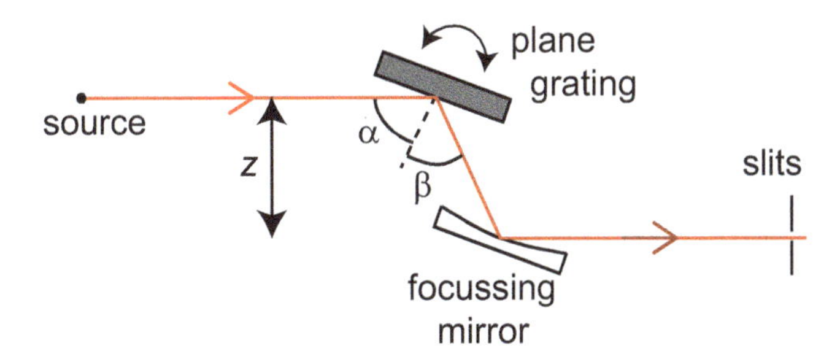

Figure 2.20. Schematic diagram showing a side view of the minimal components required for a grazing incident plane grating monochromator.

to be closed off when it is not required. As a consequence, the radiation arriving at a monochromator in the absence of any pre-focussing is essentially parallel, so in principle it is possible to use the small source size and the effective entrance slit of the monochromator as shown in figure 2.20. In practice, however, there is almost invariably some beamline pre-focussing optics, which then necessitates the introduction of a real entrance slit and some form of focussing mirror in the monochromator before the beam reaches the grating. Selection of output radiation of the required wavelength then simply requires rotation of the grating, changing both α and β, this radiation being focussed at the exit slit. The spectral resolution is determined by the degree of dispersion of the grating and the width of the exit slit.

In this particularly simple design, the post-grating focussing mirror remains fixed and the only moving part is the rotation of the grating. An important feature is that the direction of the output beam is parallel to that of the incident radiation, so the essentially linear characteristics of the beamline in its delivery of radiation to a sample chamber is retained. Indeed, particularly for surface science experiments involving a heavy UHV sample chamber, it is important that the sample position can remain fixed as the wavelength is scanned. This requirement is satisfied by this basic design. Notice, though, that there is a (constant) offset in height of the two beams. If necessary, this can be compensated by a pair of plane mirrors with the same vertical offset to the grating and focussing mirror. Of course, each grazing incidence optical component leads to some loss of intensity, so in general it is important to minimise the number of these components in a monochromator (and in the beamline as a whole). One way to reduce the number of these components is to replace the plane grating, and at least one of the focussing mirrors, by a spherical or toroidal grating, which can both disperse and focus the output radiation. A toroidal grating monochromator (or TGM) can lead to a particularly simple design: one, which operated successfully at the original BESSY facility in Berlin from the 1980s over a wide spectral range from 180 to 1150 eV, used the source as the virtual entrance slit and thus contained no other mirrors (Dietz *et al* 1985). Far more commonly, however, the alternative to a PGM is a spherical grating monochromator (or SGM).

One complication in designing a monochromator based on a spherical grating is that as one rotates the grating to select different output wavelengths, the focal properties change. This is, of course, also true for a spherical mirror: rotating the mirror changes the incidence angle and this changes the focal length and the reflection direction, as discussion in section 2.5. The 'focussing' properties of a spherical grating are defined by the Rowland circle, a circle whose radius is equal to the radius of curvature of the mirror, drawn tangential to the mirror, as shown in figure 2.21(a). A source at a position on this circle leads to a 'focus', i.e., a nominally aberration-free single-wavelength spot, at a position that also lies on this circle. Rotating the grating then also rotates the Rowland circle and is no longer consistent with fixed entrance and exit slit positions.

As an example of a solution to this problem, figure 2.21 shows a design implemented by Schwarzkopf *et al* (1998), which involves lateral movement of the grating and a plane mirror along the optical axis as the grating is rotated to select different wavelengths. Just two positions of these components shown in the figure illustrate how the virtual exit slit remains on the (rotating) Rowland circle while the plane mirror transfers this to the fixed real exit slit.

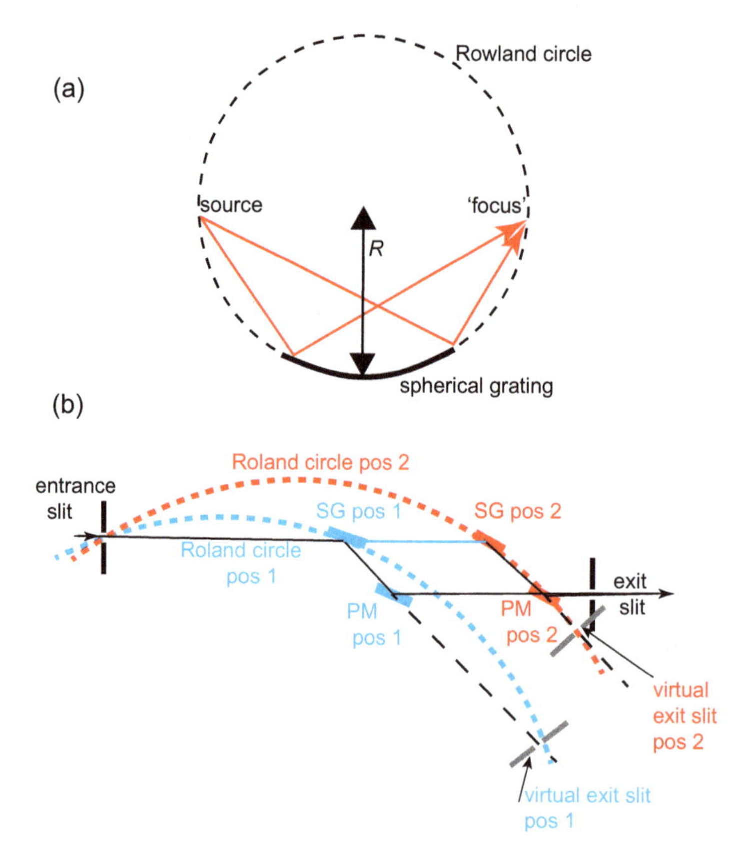

Figure 2.21. (a) Rowland circle showing the focus requirements of a spherical grating. (b) One design of a grazing incidence spherical grating monochromator designed to ensure a constant entrance and exit slit position by translating a plane mirror (PM) as the spherical grating (SG) is rotated. Just two relative positions are shown.

The resolution of all these grating monochromators is determined by the size of the entrance and exit slits and the degree of dispersion of the grating, which is defined by the periodicity or 'line spacing', d, of the grating. The ultimate resolution is typically quoted as a resolving power, $h\upsilon/\Delta h\upsilon$ (or, equivalently $\lambda/\Delta\lambda$), which is commonly approximately constant over much of the spectral range of an instrument with a particular grating, and in the extreme-ultraviolet/SXR range covering photon energies from ~50 to 100 eV can be in the range ~1000–5000. Of course, the dispersion is ultimately limited by the smallest line spacing that can be achieved in the grating manufacture. Typical widely used reflection gratings have ~1000 lines per millimetre (d = 10 μm) but may have 2–3 times smaller spacing. The highest-resolution monochromators working in this energy range, for resonant inelastic X-ray scattering spectroscopy (or RIXS), have the potential to deliver ~10 meV resolution at 1000 eV (corresponding to a resolving power of 10^5).

At higher photon energies in the hard X-ray range, typically from 3 to 30 keV or higher, with correspondingly shorter wavelengths, the grating periodicity needs to be much smaller, a requirement readily satisfied using crystalline solids. Much the most widely used crystal is silicon, and specifically the (111) Bragg 'reflection'. Of course, crystalline solids are three-dimensionally periodic, rather than the one-dimensional periodicity of manufactured diffraction gratings, and this does lead to some very significant differences in the properties of grating and crystal monochromators. To derive the condition for X-ray diffraction from a three-dimensionally periodic crystalline solid, one must identify the conditions under which scattering ('diffraction') from each atomic basis in the primitive unit cell of the crystal, defined by the primitive translation vectors **a**, **b** and **c**, leads to coherent interference. This leads to the three Laue conditions:

$$\mathbf{a}.\Delta\mathbf{k} = 2\pi h \ \mathbf{b}.\Delta\mathbf{k} = 2\pi k \ \mathbf{c}.\Delta\mathbf{k} = 2\pi l,$$

where h, k and l are integers, and

$$\Delta\mathbf{k} = \mathbf{k}_{\text{out}} - \mathbf{k}_{\text{in}}$$

is the change in the wavevector between the incident and diffracted X-rays, generally referred to as the scattering vector. This leads to the simplified Bragg's law in which the scattering is envisaged as being from planes of atoms, shown in figure 2.22, leading to

$$n\lambda = 2d \sin\theta,$$

where λ is the wavelength, d is the spacing of the scattering planes and θ is the grazing incidence angle relative to these scattering planes. Clearly the situation is quite different from that of scattering from a diffraction grating. Although the crystal is (two-dimensionally) periodic in the surface, like the one-dimensional periodicity of the grating, the diffraction condition of Bragg's law relates to the effect of the periodicity perpendicular to the surface. Of course, this law takes account of the full three-dimensional periodicity through the Laue conditions, but while the diffraction grating leads to scattering of many different wavelengths, dispersed in angle, the crystalline diffraction leads to scattering of a single

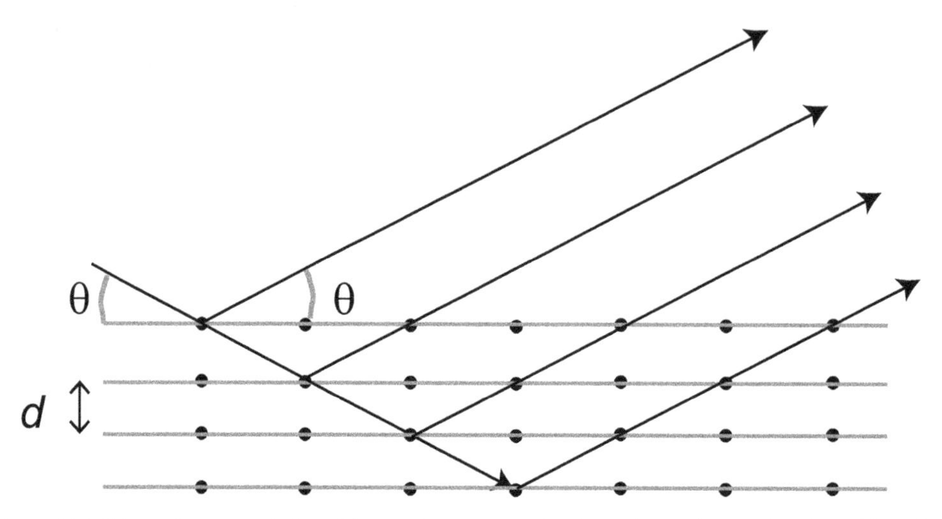

Figure 2.22. Schematic representation of Bragg's law as scattering from planes of atoms for X-ray diffraction from a crystalline solid.

wavelength in a single direction (though under some conditions diffraction from different sets of scattering planes may lead to additional diffracted beams at different specific energies and directions).

Evidently, a single crystal reflection can thus provide monochromatic radiation, but rotating the crystal to deliver different wavelengths leads to the direction of the monochromatic output beam changing, an arrangement not compatible with fixed entrance and exit slits. This can be easily overcome by using a pair of identical crystals. Indeed, a common early practice was to use a single 'channel-cut' silicon crystal such that the two diffracting crystals are part of the same large single crystal, shown in figures 2.23(a) and (b). While conceptually simple, this approach has two disadvantages. First, rotation about the centre of the crystal to select the appropriate wavelength of emitted radiation causes the height of the outgoing beam to change as the selected wavelength is changed. Secondly, the much higher heat load on the part of the crystal exposed to the incident 'white' beam can lead to a different lattice parameter for this part of the crystal, so the two components no longer match the Bragg condition for the same wavelength. This second problem, of course, has become more severe as sources have developed from bending magnets to wigglers and undulators. By separating the two component crystals, now rotated about their individual centres, the height change is avoided, while small adjustments in their relative angles can correct for any temperature differences. The development of extremely precise stepper motors has made this solution routine and fully under computer control.

The spectral resolution of such a double-crystal monochromator is determined by two factors that influence the 'rocking curve width' of the Bragg reflection. One of these is the crystalline perfection, which is the reason why the crystal that is most commonly used is silicon, a material that can be grown to be almost perfect and dislocation-free. The second factor is the extinction depth of the X-rays in the crystal. The finite penetration of the X-rays into the crystal, due not only to the

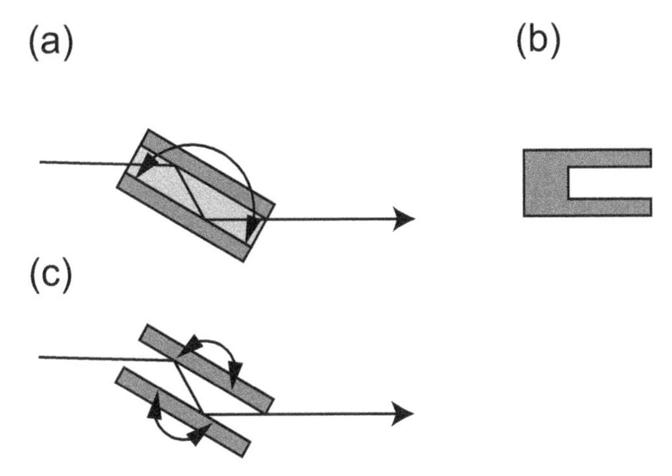

Figure 2.23. Schematic diagram of the key component of a double-crystal X-ray monochromator. Panels (a) and (c) show the incident beam trajectory through a channel-cut single crystal and the alternative arrangement of two separate crystals. Panel (b) shows the end-on shape of the channel-cut crystal.

effects of absorption but also to the elastic scattering out of the crystal, means that the intensity of the Bragg reflection as a function of scattering angle is not a delta function, as implied by the derivation of the Laue conditions for an infinite perfect crystal, but has a finite width due to this finite penetration. The origin and implication of these effects is described more fully in section 4.2, in the introduction to the structural technique of X-ray standing waves (or XSW). Typically, the resolving power of a Si(111) double-crystal monochromator is $\sim 10^4$. Higher resolution can be achieved using different reflection orders and geometries. The lower limit of the spectral range of a monochromator based on Si(111) Bragg reflections is determined by the *d* spacing of these scattering planes. This value is 3.135 Å so the ultimate limiting wavelength is 6.27 Å, corresponding to a photon energy of 1979 eV, but this would correspond to an incidence angle of 90°; evidently, a somewhat smaller angle (and thus a higher minimum photon energy) is required for operation of the double-crystal monochromator of figure 2.23. With grating monochromators typically being used for photon energies below ~1 keV, the energy range of ~1–2 keV does fall between the optimum ranges of grating and crystal monochromators. Alternative crystals with larger *d* spacings can be used to achieve lower photon energies from a crystal monochromator, although in general their degree of perfection or robustness is inferior to Si.

References

Dietz E, Braun W, Bradshaw A M and Johnson R L 1985 A high flux toroidal grating monochromator for the soft X-ray region *Nucl. Instrum. Methods* A **239** 359–65

Finetti P, Holland D M P, Latimer C J and Binns C 2004 Polarisation analysis of VUV synchrotron radiation emitted from a bending magnet source in the energy range 20–50 eV: a comparison between measurements and theoretical predictions *Nucl. Instrum. Methods* B **215** 565–76

Henke B L, Gullikson E M and Davis J C 1993 X-ray interactions: photoabsorption, scattering, transmission, and reflection at E = 50–30,000 eV, Z = 1–92 *At. Data Nucl. Data Tables* **54** 181–342

MacDonald C A 2011 Focusing polycapillary optics and their applications *X-ray Opt. Instrum.* **2010** 867049

Sasaki S 1994 Analyses for a planar variably-polarizing undulator *Nucl. Instrum. Methods* A **347** 83–6

Schwarzkopf O, Eggenstein F and Flechsig *et al* 1998 High-resolution constant length Rowland circle monochromator at BESSY *Rev. Sci. Instrum.* **69** 3789–93

Schwinger J 1949 On the classical radiation of accelerated electrons *Phys. Rev.* **75** 1912–25

Tanaka T and Kitamura H 2001 SPECTRA: a synchrotron radiation calculation code *J. Synchrotron Rad* **8** 1221–8

Chapter 3

Photoemission and the electronic structure of surfaces

The range of methods exploiting synchrotron radiation to determine the electronic structure of surfaces is described, notably based on photoemission and photoabsorption. Higher-resolution core-level photoemission provides spectral fingerprints of 'chemical shifts' associated with the different valence electronic environment of the emitter atoms, while access to a wide range of photon energies allows the technique to be extended to sub-surface phenomena and to surfaces in the presence of reactant gases. Angle-resolved photoemission at different photon energies allows direct mapping of spin-state-specific occupied states of near-surface electronic band structure, aided by control over the state of polarisation of the incident radiation. Photoabsorption as a function of photon energy above absorption edges provides complementary information on unoccupied valence states, but using circularly and linearly polarised radiation also provides detailed information on surface magnetism and ferroelectricity.

3.1 Introduction

One of the core challenges of surface science is to understand the electronic properties of surfaces and their interaction with other atoms, molecules and materials. At the most fundamental level, a surface breaks the three-dimensional periodicity of a crystalline solid, leading to modification of the electronic structure of the outermost atomic layers and the existence of surface-localised states with energies that may fall in a band gap of the underlying solid. Interaction with the surface by adsorbed atoms and molecules, and the formation of solid–solid interfaces, leads to modification of the electronic structure of all these components, relevant to both the physics and chemistry of the resulting surface or interface. Photoemission is the most widely used technique to investigate these properties and is well-established in standard surface science laboratories. Traditionally, these

doi:10.1088/978-0-7503-3847-9ch3

experiments have been performed by 'line' sources that deliver monochromatic radiation in either the vacuum ultraviolet (VUV, most typically using a He I discharge emission at 21.2 eV) or X-ray energy ranges (usually Al or Mg Kα at 1486.6 and 1253.6 eV, respectively) leading to the techniques of ultraviolet photoelectron spectroscopy (UPS) and X-ray photoelectron spectroscopy (XPS). The availability of a continuous range of photon energies at high intensity from synchrotron radiation blurs these boundaries, but there remains a distinction, generally to use lower photon energies (typically up to a few tens of eV) to obtain the most detailed information on valence electron states, and higher photon energies for core-level photoemission. Of course, core-level photoemission requires photon energies higher than the corresponding binding energies of the states of interest, many of which have values of hundreds of eV up to a few keV. The ability not only to select different photon energies, but also to be able to vary them easily over both narrow and wide ranges makes it possible to exploit a number photoemission and photoabsorption techniques not possible with sources of fixed energy, such as X-ray absorption near edge structure (XANES) or near-edge X-ray absorption fine structure (NEXAFS), and resonant photoemission.

3.2 Energy conservation and core-level photoemission

Figure 3.1 provides a reminder of the implications of energy conservation in all photoemission experiments. Photoemission involves the transfer of all the energy of each interacting photon, $h\nu$, to a bound-state electron, producing an emitted photoelectron with a kinetic energy,

$$E_{\mathrm{kin}} = h\nu - E_{\mathrm{B}}, \tag{3.1}$$

where E_{B} is the binding energy of the initial state. This is the basic Einstein relationship, which led to the award of the 1915 Nobel Prize. In this simple expression the kinetic energy and the binding energy must refer to the same energy level (e.g., the vacuum level or the Fermi level), and E_{B} is strictly the binding energy as measured by photoemission, which may differ from the one-electron ground-state energy, as discussed in more detail in this section.

The photoemission process must also conserve momentum, but at the typical energies used in most photoemission experiments the photon momentum is essentially negligible relative to the electron momentum at a similar energy. In the case of photoemission from a core level of a free atom, the momentum of the outgoing photoelectron is matched by an equal and opposite recoil momentum of the atom, but the huge difference in mass of the atom and the photoelectron (a ratio of typically 10^4–10^5) means this has minimal impact on the photoelectron kinetic energy. In the case of photoemission from an itinerant electronic valence state of a crystalline solid, however, momentum conservation must be achieved by the solid recoiling by a reciprocal lattice vector, **G**; the implications of this are discussed in the context of angle-resolved photoelectron spectroscopy (ARPES) in section 3.5.

Core-level photoemission is generally referred to by the standard laboratory technique name of XPS. Using synchrotron radiation rather than a standard

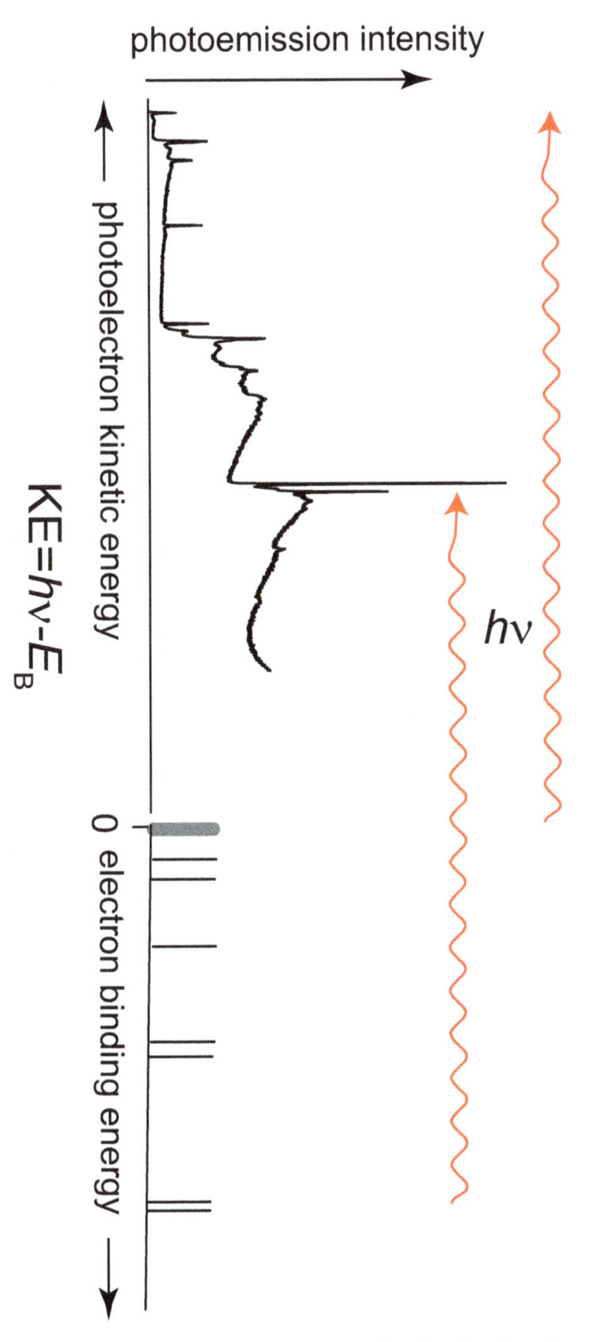

Figure 3.1. Energy-level diagram of a photoemission experiment showing the effect of transferring the energy of a photon, $h\nu$, to electrons in different bound states, E_B, leading to a photoelectron energy spectrum with peaks in kinetic energy $E_{kin} = h\nu - E_B$.

laboratory Kα X-ray source, it is sometimes referred to as SXPS, although this can be ambiguous as the 'S' may stand for 'synchrotron' or 'soft'; many synchrotron XPS studies of surface use relatively 'soft' X-ray energies. In its simplest form in surface science, it can be used to determine the elemental composition of a surface, exploiting the characteristic core-level binding energies of the different elements. However, the exact values of the photoelectron binding energy, E_B, as defined by equation (3.1), display small 'chemical shifts' associated with the local electronic or chemical bonding environment of the emitting atom. Indeed, XPS was originally known as electron spectroscopy for chemical analysis (ESCA), stressing the chemical state sensitivity as well as the elemental sensitivity of the technique; this technique was developed by K Siegbahn, for which he was awarded the Nobel Prize in Physics in 1981. The main benefits of using synchrotron radiation rather than fixed-energy Kα 'line' sources arise from the ability to select freely the photon energy. In particular, by using a photon energy only a few tens of eV above the nominal binding energy of the core levels of interest, the reduced inelastic scattering mean free path and associated sampling depth leads to enhanced surface specificity. A further advantage of low photoelectron kinetic energies is the ability to operate the photoelectron spectrometer at higher spectral resolution; of course, the highest-resolution spectra are obtained when both the monochromator of the incident radiation and the photoelectron spectrometer, each with roughly constant resolving power ($E/\Delta E$), are operated at low energies, favouring studies of core states with relatively low photoelectron binding energies using soft X-rays. However, synchrotron radiation also offers the possibility to perform photoemission experiments at much higher photon energies (commonly referred to as 'hard' X-ray photoelectron spectroscopy, or HAXPES), the higher resulting sampling depth providing access to properties of buried interfaces and more bulk-like photoemission spectra, as described more fully below.

While the original applications of the ESCA technique recognised that the measurements were, at least to some extent, surface specific (with sampling depths of ~100 Å being cited), chemical shifts in the photoelectron binding energies were generally attributed to changes in the one-electron binding energy due to different ionic charge states of the atoms in the compounds being studied. Of course, like all spectroscopies, PES provides a measurement of the energy difference between the initial ground state and an excited state, this final state corresponding to a core-ionised atom (or ion) and a photoelectron escaping into the continuum of unoccupied states. A convenient starting point in understanding the impact of this in more detail is to start by ignoring the impact of the final-state effect, in which case the quantity measured in photoemission, E_B is equated to the true one-electron binding energy. This implicitly assumes that after removing the electron from the initial core-level ground state all the electrons in the other occupied states remain at their original ground-state energies—the 'frozen-orbital' picture. This Koopman's energy, E_{Koo}, is never observed in practice. The creation of the core hole means that the remaining electrons that are more weakly bound to the emitter atom now see an increased effective nuclear charge and so become more strongly bound in the atom than before the removal of the core electron. For photoemission from a free atom this means that the photoelectron kinetic energy is increased from an initial notional value of $h\nu - E_{\mathrm{Koo}}$ to $h\nu - E_{\mathrm{Koo}} + E_a$, where E_a is the intraatomic relaxation energy. The resulting measured photoelectron binding energy is thus less than the Koopman's energy, $E_{\mathrm{B}} = E_{\mathrm{Koo}} - E_a$. Notice that this description assumes that the photoemission process is sufficiently slow that the emitter atom is fully relaxed before the escaping photoelectron ceases to 'feel the presence of' the core-ionised emitter. This is known as the adiabatic approximation. In practice, it is quite likely that the photoemission process is more rapid than this, so at the time that the photoelectron escapes, the core-ionised emitter may still be in an excited state, with some remaining electrons of the emitter in excited bound states or even excited into the continuum (thereby leaving a multiply ionised emitter atom). Assuming that the photoelectron leaves the emitter in this excited state is known as the 'sudden' approximation. As a result, the escaping photoelectrons may occur with several different values of the kinetic energy, leading to a photoelectron energy spectrum comprising a highest-energy peak corresponding to emission from the fully relaxed emitter and 'satellite' peaks at lower kinetic energies corresponding to emission from the emitter in different excited states. In this case, of emission from a free atom, satellite peaks due to different discrete excited states are known as 'shake-up', while a feature due to excitations into the continuum is known as 'shake-off'. In applications of (S)XPS to surfaces. of course, one is not studying free atoms in the gas phase, but rather atoms and molecules on or in a solid (or liquid) surface. The fact that the emitter atom is then bonded to other atoms means that the photoelectron binding energy is also influenced by interatomic relaxation, and the binding energy of shallowly bound electrons on adjacent atoms are also expected to relax due to the nearby presence of the positively charged nearby emitter atom. This results in an additional interatomic or extra-atomic relaxation energy, E_r, so the measured photoelectron binding energy is

$$E_B = E_{\mathrm{Koo}} - E_a - E_r.$$

One important aspect of the extra-atomic relaxation energy in this equation is that it means that there are two contributions to the 'chemical shifts' in the measured binding energy, E_{B}, namely, the change in the initial-state energy, but also the change in the final-state energy. Experimentally, there is no simple general way of separating these contributions that jointly contribute to the measured chemical shift. In the case of XPS from gas-phase molecules, however, the results shown in figure 3.2 indicate that the dominant component is the change in the initial-state energy.

Specifically, this comparison of C 1s photoelectron binding energies in a range of carbon-containing gas-phase molecules with calculated frozen-orbital energies shows all values lie close to a straight line of unity gradient. The implication is that in these cases the observed chemical shifts in the photoemission experiments are consistent with the dominant influence being the change in the initial-state energy rather than changes in the extra-atomic relaxation energy that is not accounted for in these calculations. Notice, though, that all experimental values are approximately 15 eV smaller than the absolute calculated values, a difference that may be attributed to intraatomic relaxation.

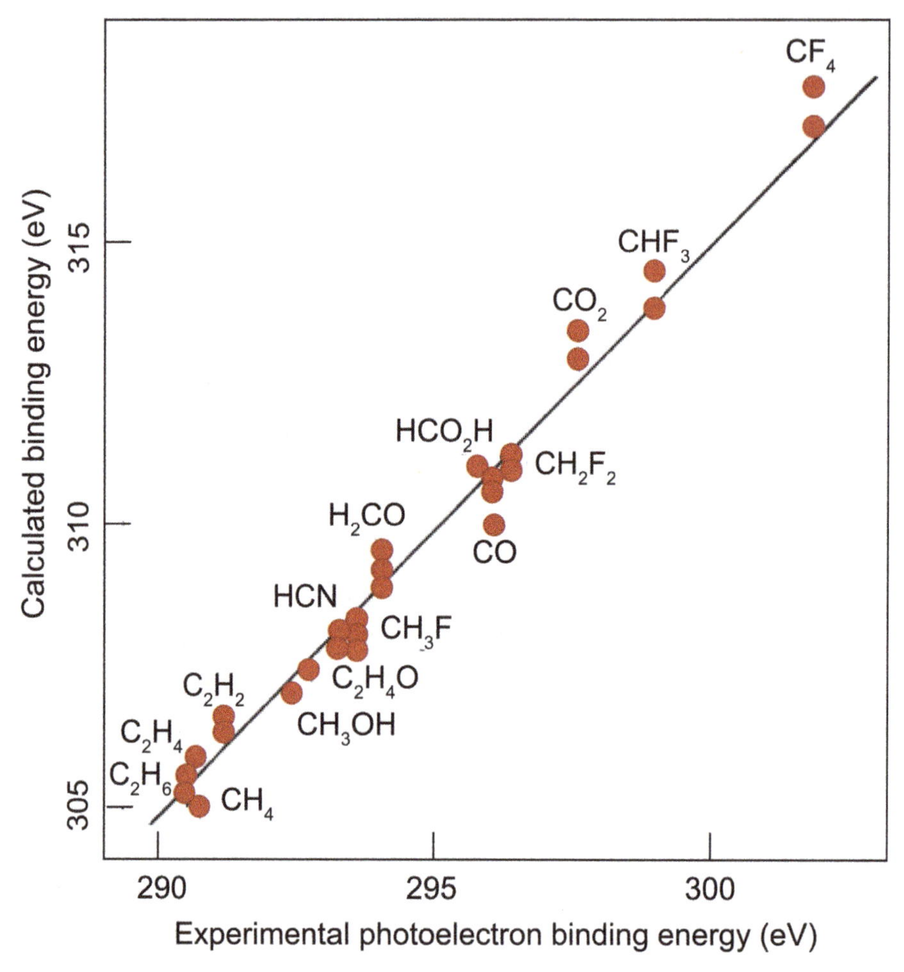

Figure 3.2. Comparison of experimentally determined C 1s photoelectron binding energies with calculated frozen-orbital energies for a range of carbon-containing molecules. Based on data presented by Shirley (1973).

Indeed, dominance of the role of the initial-state energy difference as the source of experimentally observed chemical shifts has frequently been assumed to be the case in traditional ESCA studies of solid compounds, with different chemically shifted components being assigned to different charge states of the atom. Figure 3.3 shows an example of this for V 2p and Ti 2p SXP spectra recorded from a thin film of a mixed V-Ti oxide formed by V deposition onto a rutile $TiO_2(110)$ surface. In bulk stoichiometric TiO_2 the Ti ions are generally believed to have a 4+ charge and the dominant Ti 2p and V 2p peaks in the mixed oxide have photoelectron binding energies associated with these states in the bulk oxide. A shift of ~1 eV to lower binding energy is attributed to 3+ species, while a similar increase in the photoelectron binding energy is attributed to a 5+ species. As expected, emission from both the $2p_{1/2}$ and $2p_{3/2}$ components show the same chemical shifts.

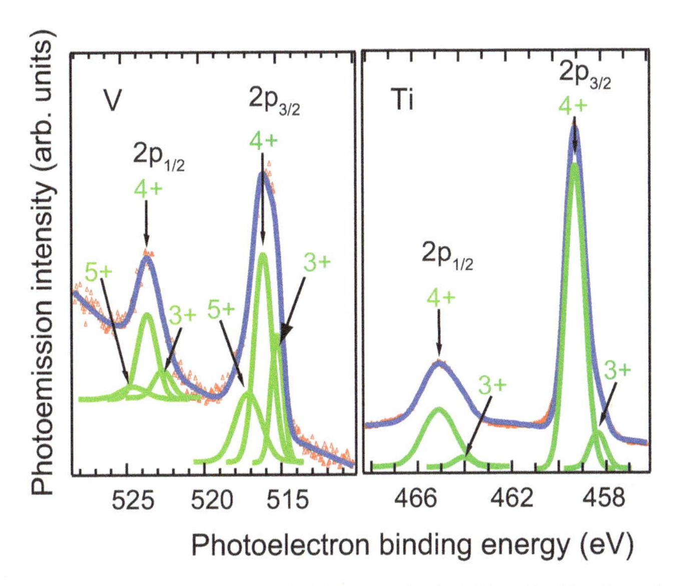

Figure 3.3. V 2p and Ti 2p SXP spectra recorded from a mixed V-Ti oxide film formed on an TiO_2(110) surface by deposition of V and subsequent annealing. The experimental data are shown as red triangles, while the blue lines show the fit to these data based on the chemically shifted components shown in green. Reprinted from Duncan *et al* (2014), copyright (2014), with permission from Elsevier.

In the case of atoms on or at surfaces this generalisation regarding the relative importance of initial- and final-state contributions to XPS chemical shifts is no longer valid. Early calculations by Williams and Lang (1977) for oxygen (O), sodium, silicon (Si) and chlorine adsorbed at the surface of a jellium model, to represent the influence of the free electrons in an aluminium (Al) surface, showed significant differences in these two contributions for these four model systems. In general, it remains true that XPS chemical shifts are simply regarded as a spectral 'fingerprint' of different bonding situations. However, using modern density functional theory (DFT) it is now possible to calculate quite accurately the chemical shifts (though not the absolute binding energies) for specific atoms as atomic adsorbates and in adsorbed molecules on specific surfaces, providing a more quantitative identification of the states associated with different photoelectron binding energies. The application of DFT calculations to the chemical shifts associated with oxidation of Ag (Grönbeck *et al* 2012) provide a particularly striking example of how interpretations based on initial-state charge transfer can be very misleading. Because O is electronegative, oxidation of a metal surface typically leads to an increase in the photoelectron binding energy of emission from a core level of the metal atoms (a positive core-level shift relative to the metal). This initial-state effect is consistent with the Ti and V XP spectra of figure 3.3: states assigned to a higher positive charge state have larger positive chemical shifts. However, oxidation of Ag is found to lead to a negative chemical shift in Ag 3d photoemission. DFT calculations comparing the predicted core-level shifts following oxidation of Pd and Ag show that, while the expected positive shift is predicted (and observed) for O on

Pd(111), O on Ag(111) leads to a negative shift. This is despite the fact that the calculations indicate that both Pd and Ag atoms do become more positively charged when adjacent to O atoms, yet this charge transfer leads to a significant change in the Pd core potential but no significant change in the Ag core potential. The conclusion is that while the O-induced chemical shift of the Pd 3d emission on Pd(111) can be attributed to an initial-state effect, the opposite shift of the Ag 3d emission from O adsorption on Ag(111) is attributed to a final-state relaxation effect. The difference in screening (relaxation) effect in the two surfaces is related to the largely *s*-character of the Ag valence states as opposed to the *d*-character of the Pd valence states.

Notice that an important aspect of this example is that the associated experiments require that the emission of photoelectrons from the surface atoms of a metal surface can be distinguished from emission of the same atoms in the underlying bulk, i.e., that there is a 'surface core-level shift' (SCLS) in the emission from the metal atoms. Of course, the outermost layer of atoms at a metal surface do experience a different electronic environment from those of the underlying bulk, due to the reduced number of neighbouring metal atoms, but also due to the decay of the valence charge density at the surface, so the existence of such an effect is not surprising. However, the ability to observe and exploit this effect depends on the size of this shift, the relative intensities of the emission from the surface and bulk layers, and the spectral width of the core-level photoemission peak that depends both on the instrumental (monochromator and electron spectrometer) resolution and on the intrinsic linewidth of the peak. In particular, emission from more localised states of *f* or *d* orbital character generally leads to sharper peaks. Nevertheless, surface core-level shifts have been observed from Al surfaces in Al 2p photoemission, although on the Al(111) surface the size of the shift was found to be only 27 meV (Borg *et al* 2004). Investigations of SCLS and the chemical shifts associated with adsorbate bonding to outermost substrate atoms benefit particularly strongly from the ability offered by synchrotron radiation to tune the photon energy such that the photoelectron energy is of order 50 eV. This ensures the highest degree of surface specificity due to the reduced sampling depth (corresponding to the broad minimum in the energy dependence of the inelastic scattering mean free path; see figure 1.2), but also allows photoelectron spectrometers with a fixed resolving power, $E/\Delta E$, to operate at particularly high spectral resolution. This enhanced surface specificity is less relevant in studies of core-level photoemission from adsorbates, their constituent atoms being present only at the surface, although a reduced sampling depth does decrease the intensity of the inelastically scattered background of substrate emission, thereby improving the signal-to-background ratio of the adsorbate emission.

An investigation of carbon monoxide adsorption on Rh(111) provides a nice example of how a combination of core-level shifts in emission from the outermost metal atoms and the atoms within the adsorbed molecule can lead to qualitative structural information. A reproduction of the key spectra recorded by Beutler *et al* (1998) is shown in figure 3.4. CO adsorption on this surface leads to several ordered phases at increasing coverage, but two of these, at CO coverages of 0.25 and 0.75 Ml, both have (2×2) unit meshes. Increasing coverage leads to increased suppression of the intensity of the SCLS Rh $3d_{5/2}$ peak of the clean-metal surface,

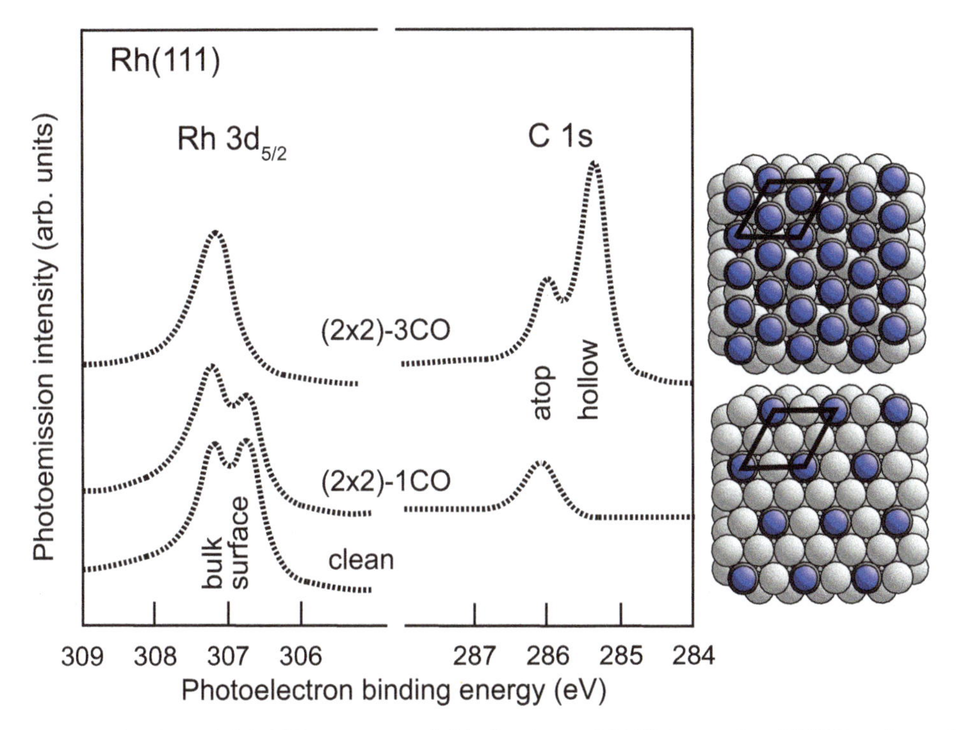

Figure 3.4. Rh $3d_{5/2}$ and C 1s SXP spectra recorded by Beutler *et al* (1998) from a clean Rh(111) surface and from two different CO adsorption phases on this surface. On the right are schematic representations of the structure of these two phases showing superimposed (2 × 2) unit meshes. Reprinted from Beutler *et al* (1998), copyright (1998), with permission of Elsevier.

while the C 1s emission shows a single peak at low coverage, joined at higher coverage by a second peak that ultimately has a significantly higher intensity. Also shown in figure 3.4 are schematic diagrams of the structural models of the two (2 × 2) phases. At a coverage of 0.25 Ml all the CO molecules in this (2 × 2)-1CO phase occupy sites atop surface rhodium (Rh) atoms, whereas in the (2 × 2)-3CO phase an additional 0.5 Ml of CO molecules occupy hollow sites. The two C 1s peaks can be attributed to emission from the atop and hollow CO species, respectively. Notice that in the (2 × 2)-1CO phase only 25% of the surface Rh atoms are bonded to CO molecules. This accounts for the fact that the Rh $5d_{5/2}$ peak attributed to (clean) surface atoms is attenuated but not fully quenched. In the (2 × 2)-3CO phase all surface Rh atoms are bonded to CO molecules, so this peak is completely suppressed. Notice, too, that if the CO molecules occupied hollow sites in the low-coverage (2 × 2) phase, 75% of the surface Rh atoms would be bonded to CO molecules, which would be expected to lead to a much stronger suppression of the intensity of the clean surface Rh $3d_{5/2}$ peak. The modest suppression of the intensity of this peak in the (2 × 2)-1CO phase not only indicates that the atop sites are preferentially occupied, but also that the SCLS effect is very localised. Of course, synchrotron radiation is not essential to resolve chemical shifts associated with CO adsorbed in different sites. This same effect was reported for CO and CO/H coadsorption on Ni(100) using a monochromated Al Kα source with a spectral resolution of 0.4 eV (Tillborg *et al* 1992). The enhanced resolution possible with monochromated soft X-rays at a synchrotron radiation source can, however, provide additional information in such studies and access a wider range of systems.

Figure 3.5 shows C 1s spectra taken with a total instrumental resolution (photons and electrons) of 0.17 eV by Smedh *et al* (2001) from the same Rh(111)(2 × 2)-3CO phase as shown (at lower resolution) in figure 3.4. The two main peaks associated with CO adsorbed in the top and hollow sites show clear shoulders at lower kinetic energy that can be assigned to the excitation of single and multiple C–O stretching vibrations, as shown by the component fit in figure 3.5. The energy of these stretching vibrations, measured in many studies using reflection–absorption infrared spectroscopy (or RAIRS) and high-resolution electron energy-loss spectroscopy (or HREELS), has long been regarded as a spectral fingerprint of different CO coordination sites at surfaces, and the different energy-loss features for the atop and hollow C 1s emission peaks of approximately 235 and 173 meV are consistent with this literature. Vibrational loss peaks are well known in core-level photoemission from gas-phase molecules, but observing this effect has proved more challenging for surface adsorption studies, though the additional information to be

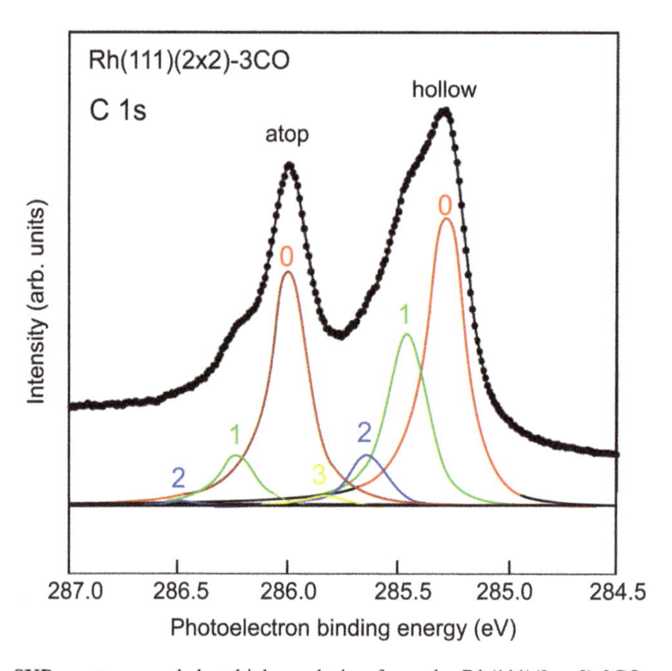

Figure 3.5. C 1s SXP spectra recorded at high resolution from the Rh(111)(2 × 2)-3CO surface phase at a photon energy of 328 eV. The fits to the experimental data include contributions, as shown, involving excitation of 0, 1, 2 and 3 C–O stretching vibrations. Reprinted from Smedh *et al* (2001), copyright (2001), with permission of Elsevier.

gained is potentially of great value in achieving a complete understanding of the SXP spectra and the assignments of the chemically shifted components.

3.2.1 'Near-ambient' pressure photoemission

Implicit in most modern surface science studies is the fact that all experiments are conducted under ultra-high vacuum (UHV) conditions. Commonly the major reason for this is the need to keep the surfaces that are being studied uncontaminated by residual gaseous species, both before and during the experiments. However, many surface science experiments are directed at gaining an improved understanding of processes, such as heterogeneous catalysis and corrosion, which involve the interaction of the surface with reactants at much higher pressures, even in some cases at pressures in excess of standard atmospheric or 'ambient' pressures. It is thus a matter of concern as to whether model studies at very low pressures of reactants do behave in the same way as reactions between the same species at 'real world' pressures. Of course, surface science techniques can be used to study (in UHV) surfaces both before and after higher-pressure reactions, but this surface analysis approach clearly does not allow the time evolution of the important surface reaction processes to be followed. True *in operando* studies clearly require analysis during reactions at more realistic pressures in order to address this 'pressure gap' in surface science. Here, we confront a second reason for the use of UHV in surface science studies: that at least some of the modern techniques of surface science *require* UHV in order to be viable. Photoemission, in particular, falls into this category; the mean free path for electrons at energies of tens to hundreds of eV in air at atmospheric pressure is only of the order of a few microns, increasing to ~1 mm at a pressure of ~1 Torr. Evidently a key requirement of an instrument capable of obtaining XP spectra from a surface while exposed to a 'near-ambient' pressure (NAP) of up to a few Torr is that the emitted electrons must travel no more than ~1 mm in this pressure. This implies that the vessel containing the sample and its elevated pressure must be separate from the vessel containing the electron spectrometer, these two vessels being linked by a small aperture (<~1 mm) through which the emitted photoelectrons must pass; this small aperture also makes it possible to maintain the large pressure difference between the two vessels by differential pumping. The earliest instruments to exploit this idea, based on laboratory X-ray sources, go back to the 1970s (e.g., Joyner *et al* 1979), but in the last decade or so an increasing number of instruments have emerged, benefitting from improvements in the design of electron optics, but also from the trend of using synchrotron radiation rather than a conventional X-ray source. The fact that the sample must be so close to the photoelectron transfer aperture means that one requires a focussed incident beam of X-rays, a requirement more easily satisfied using synchrotron radiation. Of course, the UHV of the synchrotron radiation beamline must also be separated from the elevated pressure of the sample environment, typically by a transparent window comprising a silicon nitride or Al film. The review by Trotochaud *et al* (2017) describes many of the practical considerations as well as applications in the development of NAP-XPS instruments. Figure 3.6 shows schematically the layout

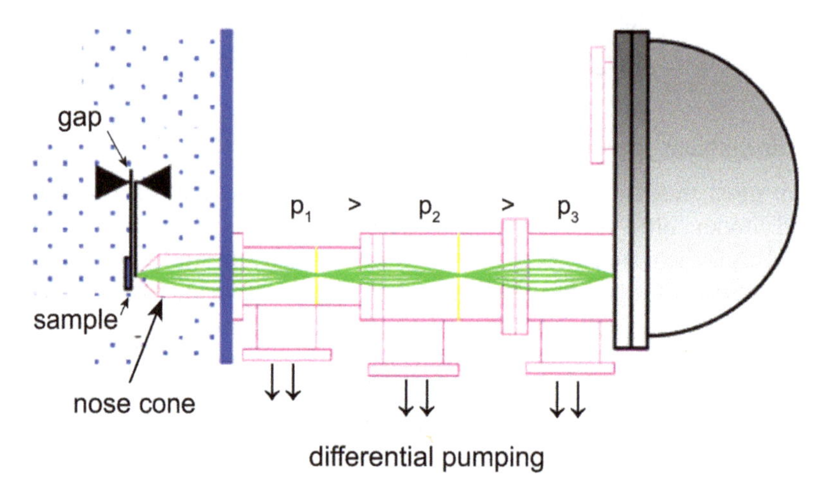

Figure 3.6. Schematic diagram showing one design of the electron spectrometer and its input lenses together with the differential pumping in a NAP-XPS instrument. Reprinted from Salmeron and Schlögl (2008), copyright (2008), with permission of Elsevier.

of a sample in its surrounding higher pressure together with the concentric hemispherical electron spectrometer (described more fully in section 3.5) and associated electron lenses and differentially pumped chambers. The use of these multiple chambers makes it possible to operate the sample environment at higher pressures. The nose cone of the analyser optics in this design is interchangeable, with several different-sized apertures and a sub-millimetre working gap from the sample.

One example of the application of NAP-XPS is to the study of surfaces under a higher partial pressure of water than is possible in conventional UHV surface science. Under standard atmospheric 'real-world' conditions, surfaces invariably have a thin covering of water, while water is a key reactant or product in many situations of corrosion or heterogeneous catalysis (as well as at electrochemical interfaces). Figure 3.7 shows an O 1s XP spectrum recorded by Yamamoto *et al* (2008) from a Cu(110) surface at 295 K under a partial pressure of 1 Torr of water. Notice that in addition to the broad feature at a binding energy of ~531–533 eV, attributable to O-containing surface species, the spectrum also shows a peak at a binding energy of ~536 eV that is from H_2O in the gas phase.

NAP XP spectra thus can provide information not only on the character of surface species during a 'high-pressure reaction', but also information on the reactant and product species in the gas phase close to the surface. As in UHV XPS studies, the ability to obtain high-resolution spectra using synchrotron radiation provides enhanced insight into the nature of the surface species present during such a reaction. In the case of figure 3.7, a study of the surface component of the O 1s spectrum as a function of temperature allows the individual component species to be identified. Notice that if water is adsorbed onto Cu(110) at low pressure under typical UHV surface science conditions it does partially dissociate, but intact water desorbs at a temperature of ~170 K. By contrast, under a partial pressure of

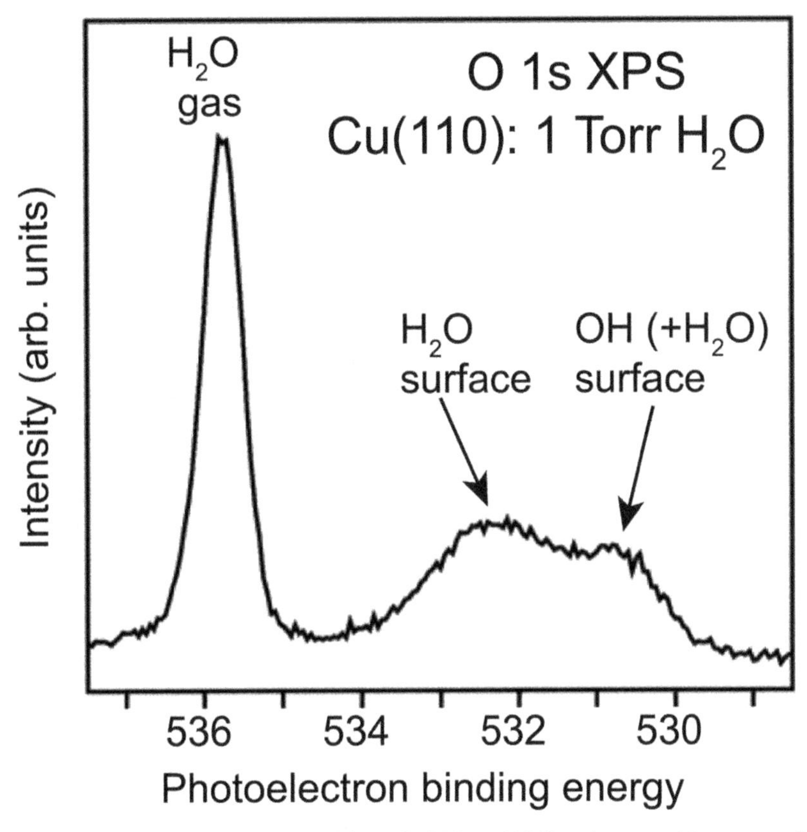

Figure 3.7. NAP XP O 1s spectrum recorded from Cu(110) at 295 K under a partial pressure of water of 1 Torr. Reproduced from Yamamoto *et al* (2008). © IOP Publishing Ltd. All rights reserved.

1 Torr of water, intact water molecules remain on the surface up to a temperature of at least 378 K.

3.3 HAXPES

While the ability to select the photon energy to yield photoelectron energies of ~50 eV ensures that the resulting photoemission spectra are highly surface specific, in studies of shallowly buried interfaces a larger sampling depth may be required, and synchrotron radiation also allows one to achieve this by selecting much higher photon energies in the (relatively) 'hard' X-ray energy range. In recent years a number of synchrotron radiation beamlines have been constructed to exploit this idea by performing 'hard XPS' (or HAXPES). These experiments are generally motivated either by the wish to study the composition and electronic structure of buried interfaces or simply to minimise the contribution of the signal arising from the near-surface region to obtain 'truly bulk' information on the electronic structure. Typically, these experiments have been performed at X-ray energies up to ~10 keV. Varying the photon energy, and thus the photoelectron energy, offers a route to non-destructive 'depth profiling', an alternative to sequentially removing surface layers by ion sputtering, or to varying the sampling depth by using different emission angles.

Performing HAXPES with 10 keV photons does present some challenges. The first of these, resolved relatively simply, arises from the need to operate the electron energy analyser at higher voltages than standard commercial instruments were designed to operate. A more fundamental instrumental challenge is the ability to achieve a high spectral resolution (in terms of both the monochromaticity of the incident radiation and the detection of the photoelectrons) at these higher energies. This is particularly relevant in measurements of the valence-band emission in order to extract information on the partial density of occupied states, but is needed for chemical state identification in core-level photoemission, for which a resolution significantly less than 1 eV is typically required. As is typically true in spectroscopy generally, there has to be a balance between resolution and signal strength; working at higher photon energies tends to degrade both of these. At a photon energy of ~10 keV, the standard Si(111) double-crystal monochromator has a resolving power of $\sim 10^4$, offering an incident radiation resolution of ~1 eV, although using higher-energy, higher-order reflections (e.g., figure 3.8) or different crystal reflections, it is possible to achieve a resolution of less than 0.1 eV; relatively standard crystal monochromators can therefore suffice for many HAXPES applications. Achieving sub-1 eV resolution in an electron spectrometer is also not particularly difficult, although high-energy photoelectrons must be retarded, prior to passing through a typical concentric hemispherical analyser at lower energies; the associated retarding optics increases the divergence, with an attendant reduction in the acceptance solid angle and thus in the detected photoemission intensity.

The extent to which HAXPES can provide sub-surface information is determined by the electron energy dependence of the effective attenuation length (EAL),

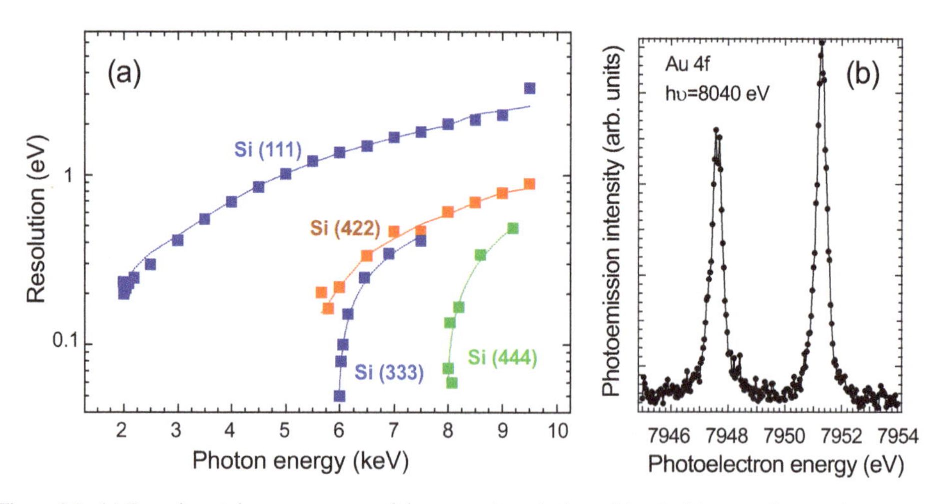

Figure 3.8. (a) Experimental measurements of the spectral resolution of the double-crystal monochromator at the KMC-1 HAXPES beamline at BESSY. Panel (b) shows the Au 4f photoemission spectra recorded at this beamline using the fourth-order of the Si(111) monochromator. Reprinted from Gorgoi *et al* (2009), copyright (2009), with permission from Elsevier.

discussed briefly in chapter 1. A computer code available from NIST (Powell and Jablonski 2011) is available to predict the appropriate value of the EAL for different materials, different energies and different experimental geometries in the photoelectron energy range up to 2 keV, appropriate to relatively standard laboratory XPS instruments operating with a high degree of surface specificity. An experimental study of the energy dependence of the EAL in gold for electron energies up to 15 keV, by Rubio-Zuazo and Castro (2011), reviews a number of previously published formulae to predict EALs at these higher energies and concludes that, for Au, EAL (nm) = 0.022 E_{kin} (eV)$^{0.627}$ provides a good fit to the data, as shown in figure 3.9. This corresponds to values of the EAL of 4.7, 7.3 and 9.4 nm at electron energies of 5, 10 and 15 keV, respectively. These values, of course, are material dependent, as may be inferred from figure 1.2. For example, Rubio-Zuazo and Castro remark that values for Al are approximately a factor of 2 larger than those of Au. Nevertheless, these values clearly do demonstrate the very significant increase in sampling depth that can be achieved in HAXPES experiments, while significant differences in EAL using X-ray energies in the range 5–15 keV highlights the potential flexibility offered by using synchrotron radiation.

HAXPES at special synchrotron radiation beamlines has proved to be effective in understanding a wide range of technologically relevant problems ranging from studies of buried interfaces in semiconductor device materials such as SiC–SiO_2 interfaces (Berens *et al* 2020) and amorphous tin-gallium oxide buffer layers in

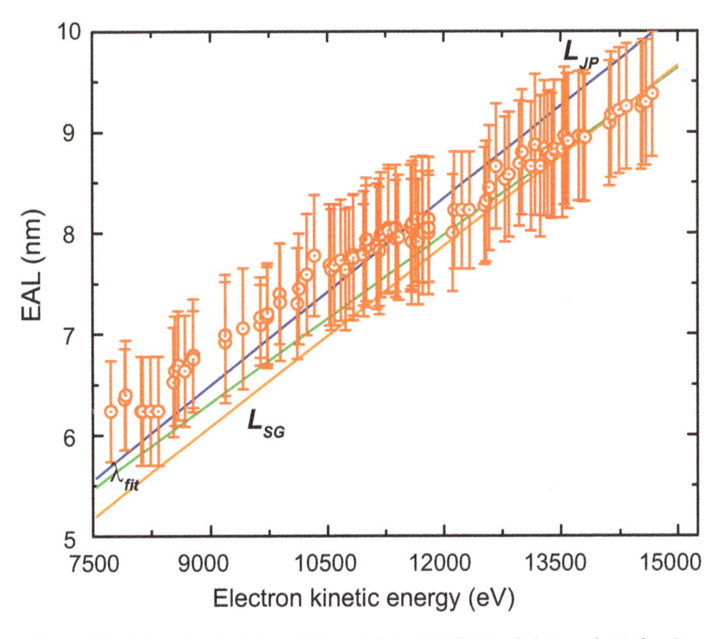

Figure 3.9. Experimentally determined values of the EAL (red individual data points) for Au as a function of electron energy, compared with the predictive formulae of Seah and Gilmore (2001), L_{SG}, Jablonski and Powell (2009), L_{JP}, and Jablonski and Tougaard (1998), λ_{fit}. Reprinted from Rubio-Zuazo and Castro (2011), copyright (2011), with permission from Elsevier.

(Ag, Cu)(In, Ga)Se_2 solar cells (Larsson *et al* 2020) to non-destructive depth profiling of passive films on stainless steel surfaces (Fredriksson *et al* 2012). A HAXPES investigation of alkali solid electrolyte interfaces (SEIs) by Gibson *et al* (2022) highlights the importance of careful control of interface formation to avoid misleading conclusions. This study sought to characterise the interface of a thin (~10 nm) lithium (Li) film on a specific argyrodite solid sulphide electrolyte (Le_6PS_5Cl) and a single graphene probe surface, using alternative methods of film deposition and preparation with and without glove-box transfers. HAXPES data were collected using incident photon energies of 2.2 and 6.6 keV, alongside soft X-ray PES (SOXPES) using photon energies chosen to give photoelectron energies of 315 eV for the peaks of interest. In Li metal the corresponding sampling depths at these energies are ~7, 19 and 1.1 nm, respectively.

Figure 3.10 shows spectra recorded in this study from the as-loaded Li_6PS_5Cl argyrodite, from 20 nm Li film evaporated on Li_6PS_5Cl *in situ*, and from 20 nm Li on Li_6PS_5Cl after transfer in an argon atmosphere. The SOXPES data evidently sample the near-surface region and show the effects of reaction with trace O, but also show S 2p intensity after the evaporation of the 20 nm Li film, attributed to reaction of mobile S^{2-} species with metallic Li. A metallic Li^0 peak is most prominent in the spectra at higher probing depths from the *in situ* evaporated film, but is absent in the spectra recorded from the sample that spent 6 hr in an Ar atmosphere in a glove box. The UHV *in situ* interface preparation clearly leads to more interpretable results. This study also explored alternative methods of depositing the Li film, finding thermal evaporation avoided the significant damage to the interface that resulted from sputtered films. In assigning the chemically shifted photoelectron binding

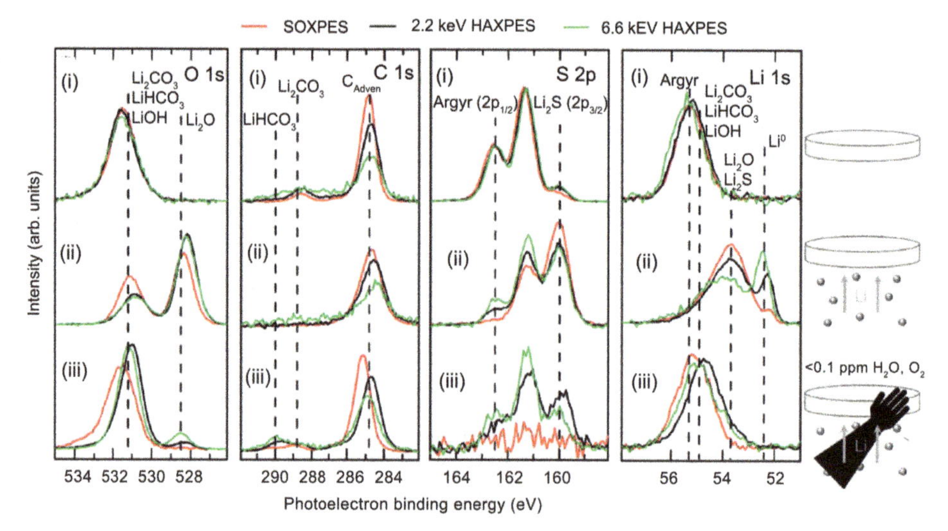

Figure 3.10. SOXPES and HAXPES O 1s, C 1s, S 2p and Li 1s spectra recorded from (i) as-loaded Li_6PS_5Cl argyrodite, (ii) 20 nm Li film evaporated on Li_6PS_5Cl *in situ* and (iii) 20 nm Li on Li_6PS_5Cl after transfer in an Ar atmosphere. Intensities normalised to same peak areas with the exception of the S 2p data in (iii). Reproduced from Gibson *et al* (2022) with permission from the Royal Society of Chemistry.

energies from low-atomic-number atoms with high kinetic energies of electrons, it is important to take account of the effect of the recoil momentum in the photoemission process. In introducing core-level photoemission in section 3.2 this effect was discounted because of the high ratio of the nuclear and electron masses. In HAXPES from low-atomic-number atoms the effect is not negligible. The approximate recoil energy shift is $\Delta E = E_{\text{kin}}(m/M)$, where m and M are the electron and nuclear masses, respectively. In the case of 6.6 keV HAXPES from Li, this leads to a recoil shift of ~0.5 eV, a significant quantity on the scale of typical chemical shifts.

Kalha *et al* (2021) presented a review of the HAXPES technique up to 2020, with a particular emphasis on the instrumental aspects and the different sources available. In this regard it is important to note that conventional sources of Kα radiation exist from target materials with higher atomic numbers than Mg and Al, the most commonly used sources in conventional laboratory-based XPS; indeed, commercial HAXPS systems are now available. Of course, these sources do not have the flexibility of continuously variable photon energies offered by synchrotron radiation.

While XPS is most widely used to exploit access to core-level photoemission and the associated chemical shifts, the most energetic photoelectrons in an XP spectrum are due to emission from the valence states, and the shape of this valence-band spectral feature can provide information on the valence-band density of states (DOS). Using standard Mg and Al Kα X-ray sources the associated photoelectron energies (<~1500 eV) must still provide a degree of surface specificity, whereas using the much higher energies of HAXPES the extent to which the valence-band photoemission reflects the properties of the surface is significantly reduced. This has led to some use of HAXPES as a means to obtaining 'true bulk' valence DOS information. This is, of course, a particularly challenging experiment, needing not only good spectral resolution for the highest photoelectron energies, but also having to combat the very low photoionisation cross-sections associated with photoemission at energies so far above the threshold for photoemission. One example of such a study is that of the transparent conducting oxide, *n*-type CdO, of interest in solar cell applications, by Mudd *et al* (2014). Obtaining a good description of the electronic properties of metal oxides of this type using DFT is challenging, so an important objective of this study was to assess the relative merits of different DFT computational approaches relative to experimental data. The valence states of this material derive from O 2s and 2p orbitals and Cd 5s, 5p and 4d orbitals, and HAXPES measurements at different photon energies offer one way to assess the relative importance of these different contributions to different parts of the valence DOS.

Figure 3.11 shows computed values of the one-electron photoionisation cross-sections for these individual valence orbitals. As is clear from the logarithmic scale, these cross-sections fall by several orders of magnitude as the photon energy is increased from a few tens of eV to several keV. This is particularly spectacular for the O 2p orbital, although it seems that the cross-section for this state is underestimated by these calculations in its contribution to the valence DOS of the oxide. Figure 3.12(a) shows the HAXPE spectra recorded at a series of different photon

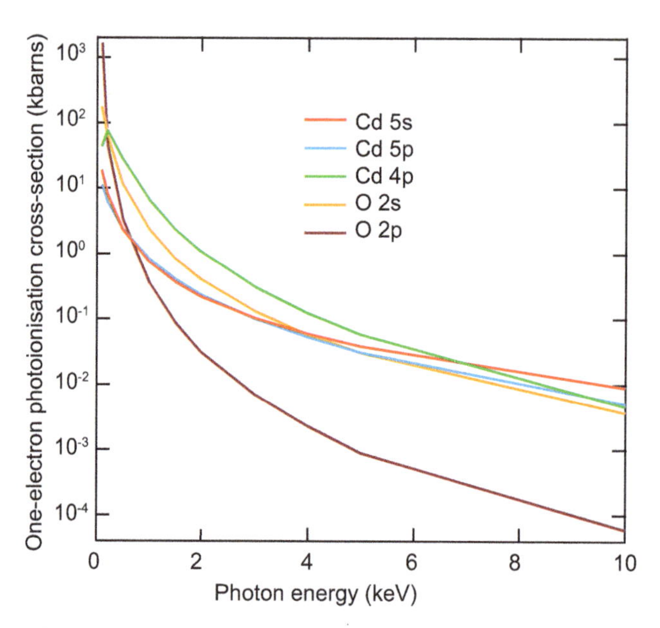

Figure 3.11. Theoretical values of the one-electron photoionisation cross-sections from Cd and O valence orbitals as a function of photon energy. Reprinted figure with permission from Mudd *et al* (2014), copyright (2014), by the American Physical Society.

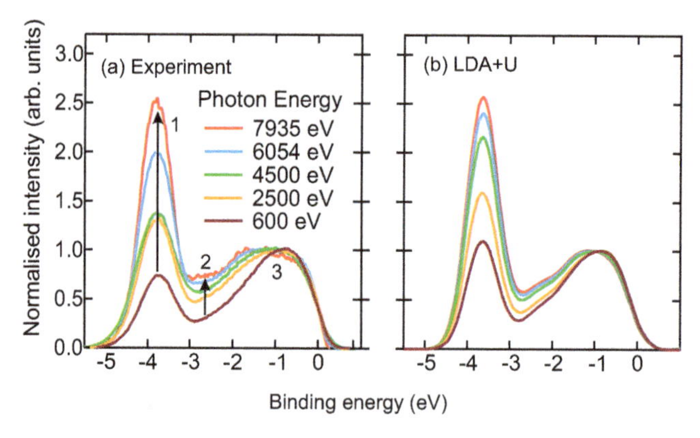

Figure 3.12. Comparison of (a) experimental HAXPE spectra from *n*-type CdO at several different photon energies with (b) theoretically generated spectra based on DFT calculations using the LDA+U functional. Reprinted figure with permission from Mudd *et al* (2014), copyright (2014), by the American Physical Society.

energies by Mudd *et al* (2014), the intensities being normalised to that of the peak labelled 3.

Clearly, as indicated by the arrow, peak 1 in the experimental valence-band spectrum increases very significantly in intensity relative to peak 3 as the photon energy increases. The intensity of the emission in the middle of band (2) also increases relative to that of peak 3. These trends are reproduced by the results of theoretical calculations. The experimental spectra were compared with theoretical spectra generated by DFT calculations using different approximations, the individual orbital contributions being broadened by a Gaussian while the intensities were scaled according to the predicted cross-sections of figure 3.11 (also corrected for the experimental geometry and the theoretically computed angular distributions). The calculations giving the best agreement with experiment were based on the LDA+U (Local Density Approximation with Hubbard U correction) DFT approach and are shown in figure 3.12(b).

3.4 Resonant photoemission

A quite different way of unscrambling the different atomic orbital contributions to the valence DOS of compounds is through the use of resonant photoemission (RPES). Specifically, this resonance occurs when the photon energy used for the photoemission coincides with the threshold energy for photoexcitation of a core-level electron to the lowest unoccupied states of the valence band, shown schematically in figure 3.13. Observing this process exploits the full control of photon energies offered by synchrotron radiation.

Figure 3.13(a) shows the standard valence-band photoemission process, while figure 3.13(b) shows photoexcitation of an electron in a core level to the lowest unoccupied state. The resulting core hole can then be refilled by an electron at the top of the valence band in an Auger-like (or autoionisation) process. If the same photon energy is used in (a) and (b), then the final states of (a) and (c) are identical, namely, a single hole in the valence band and an emitted electron with the same kinetic energy. In this case the two processes are coherent, and interference between them must occur, which in some cases leads to a strong enhancement of the intensity of the emitted electron, the resonance having a Fano line shape.

Figures 3.14 and 3.15 show the results of an application of the RPES technique to investigate the valence-band orbital contributions in a single crystal of the low-T_c (low critical temperature) ferromagnetic Kondo system $CeAgSb_2$ by Chuang *et al* (2021).

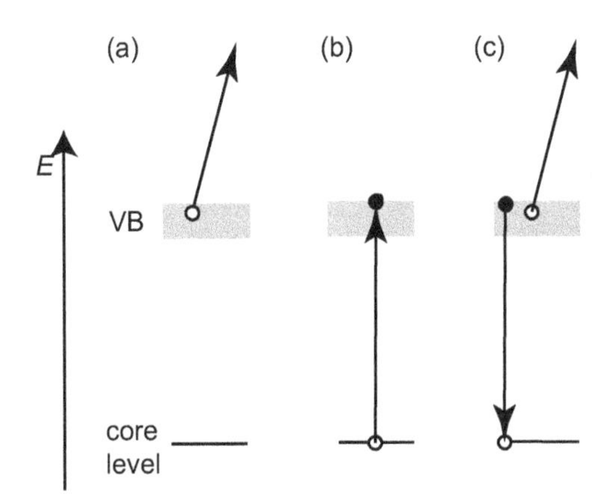

Figure 3.13. Schematic energy diagram showing the processes involved in resonant photoemission from the valence band of a sample. Panel (a) shows the conventional photoemission process, (b) photoexcitation from a core level to the lowest unoccupied state, while panel (c) shows the Auger de-excitation or autoionization event leading to the refilling of the core hole created in (b).

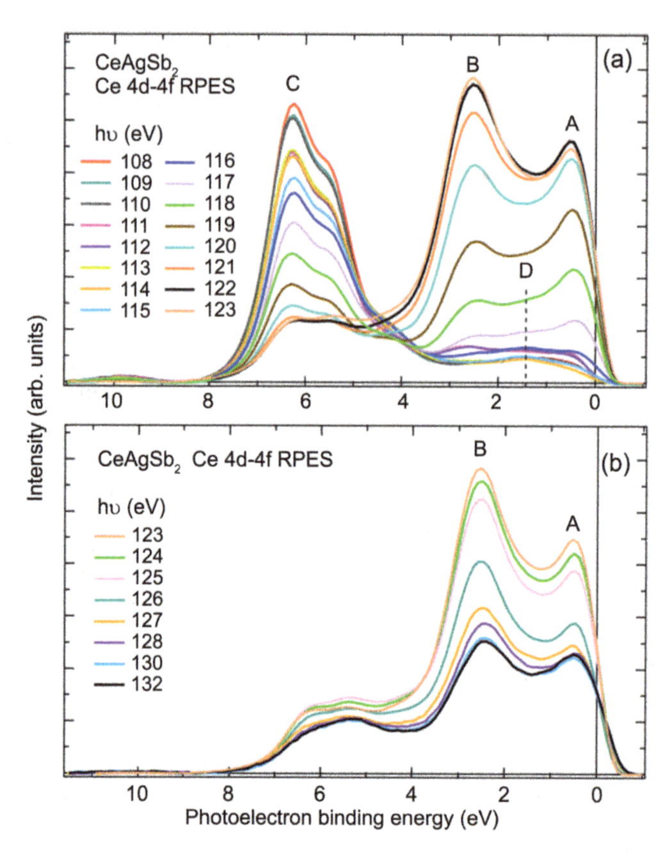

Figure 3.14. Valence-band spectra of $CeAgSb_2$ recorded at a series of photon energies scanning though the Ce 4d→4f excitation condition. Reproduced from Chuang *et al* (2021). © IOP Publishing Ltd. All rights reserved.

Clearly the intensity of the photoemission peaks labelled A and B in the binding energy range ~0–4 eV show a strong enhancement as the photon energy passes through ~121 eV, corresponding to the threshold for Ce 4d→4f excitation, indicating that states in this binding range of the valence band arise from Ce 4f orbitals. That this can clearly be attributed to RPES is shown in the constant initial state (CIS) spectra of figure 3.15, which demonstrate that the intensity enhancement of the photoemission in the A and B peaks shows a characteristic Fano line shape, as does emission from the shallow Ce 5p core level. The difference in resonance energy maximum for the A and B valence states is attributed to the dominance of bulk f^0 and surface f^1 character in the A and B peaks, respectively. Figure 3.15 also shows that the photon energy dependence of peak C is quite different, even appearing to show a weak anti-resonance. The origin of this is quite different, and attributable to a Cooper minimum in the intensity of the Ag 4d emission. The general dipole selection rule, $\Delta l = l \pm 1$, means that for emission from all states other than *s* states ($l = 0$) there are two outgoing waves, of character $l + 1$ and $l - 1$, that can interfere. The situation is particularly simple in the case of emission from a state having a principal quantum number $n = l + 2$, such as 2s, 3p, 4d, etc., for which the

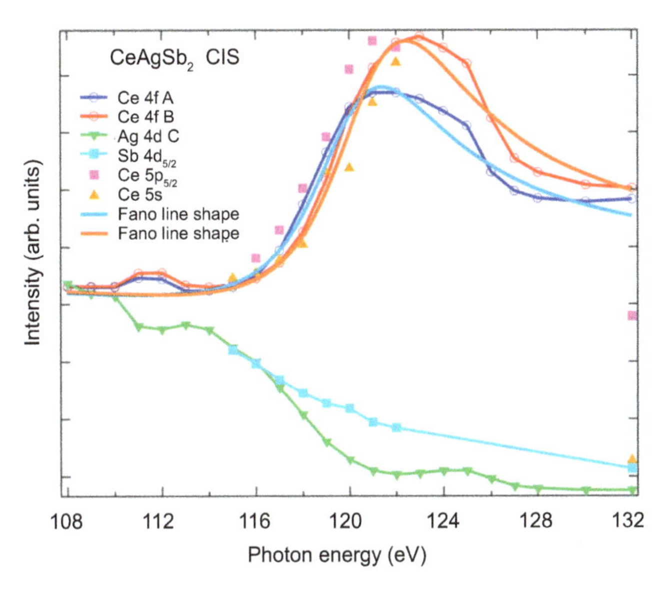

Figure 3.15. Constant initial state (CIS) spectra of different spectral components taken from the photoemission spectra of figure 3.14. Reproduced from Chuang *et al* (2021).

wavefunction has a single radial node. In this case as the photon energy increases, the matrix element for excitation to the 'up' channel ($l + 1$) changes sign, and thereby passes through zero. This leads to the total photoemission cross-section passing through a minimum. In the case of 4d emission from atomic Ag this occurs at a photon energy ~120 eV. Notice, though, that in Ag metal this Cooper minimum is shifted down to a photon energy of 60–80 eV; in $CeAgSb_2$ it appears that the Ag 4d emission is more free-atom-like, attributed to the larger Ag–Ag spacing in this compound.

3.5 ARPES: angle-resolved photoelectron spectroscopy

3.5.1 ARPES of valence-band states

So far in this chapter only the implications of energy conservation in photoemission have been discussed. It was noted that momentum must also be conserved in photoemission, but in the case of emission from core levels localised on individual atoms the recoil momentum of the emitted electron can be provided by the far more massive atomic nucleus with minimal impact of the photoelectron energy, the only property of the photoelectron that has been discussed so far. However, in using photoemission from the occupied electronic band states of a solid, it is important to recognise that these electrons have a well-defined momentum, defined by the E–k band structure of the solid, and the recoil momentum of the photoemission process must be accommodated by the solid as a reciprocal lattice vector of the solid, **G**. In the reduced-zone representation of the band structure photoemission corresponds to a typical 'vertical' or 'direct' optical transition (figure 3.16).

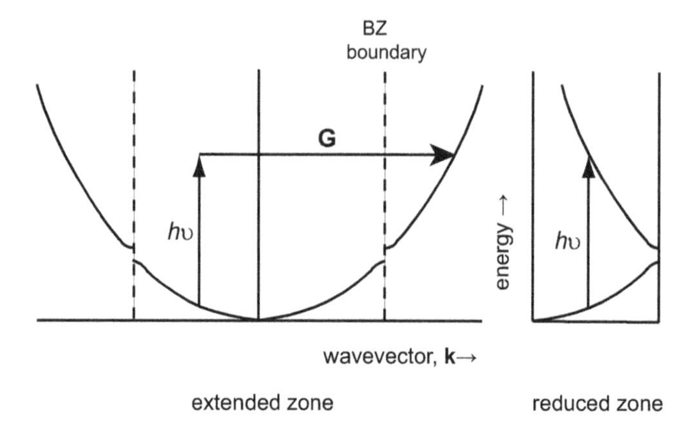

Figure 3.16. Representation of energy and momentum conservation in photoemission from a nearly-free-electron band. The near-zero photon momentum means the transition to an available final state requires a crystal momentum of a reciprocal lattice vector, **G**, to a state in the next Brillouin zone (BZ). On the left this is shown in the extended-zone scheme, on the right in the reduced-zone scheme.

However, as discussed in chapter 1, electrons emitted from a solid in the kinetic energy range of a few tens to hundreds of eV escape the surface without inelastic collisions only if they emerge from the outermost few atomic layers of the surface. Although this region of a crystalline solid retains two-dimensional periodicity parallel to the surface, it is not truly periodic perpendicular to the surface. This lack of perpendicular periodicity is a consequence of the decreasing contribution of lower atomic layers to the detected photoemission signal, even if the interlayer spacings are unchanged (which is often not the case due to surface relaxation). This means that while the component of electron momentum parallel to the surface, $\mathbf{k}_{||}$, modulo a reciprocal lattice vector, is strictly conserved in photoemission from band states, the component of the electron momentum perpendicular to the surface, $\mathbf{k}_{\perp}$, is not. In practice, however, the fact that emission from 'bulk' states is from several layers with essentially periodic spacing means that a vestige of 'smeared' $\mathbf{k}_{\perp}$ conservation is retained. These conservation rules render ARPES an exceptionally valuable tool for directly mapping the valence-band electronic structure. The basic geometry of the ARPES experiment is shown in figure 3.17.

An ARPES experiment measures the photoelectron energy, E_{kin}, of peaks in the energy spectrum corresponding to emission from occupied initial states, together with the polar angle of the detector relative to the surface normal. The value of $k_{||}$ (in Å^{-1}) corresponding to these peaks is then

$$k_{||} = 0.5123 E_{\text{kin}}{}^{1/2} \sin\theta, \tag{3.2}$$

where the kinetic energy is expressed in eV. Notice that because $k_{||}$ is conserved as the electron travels from inside to outside the crystal, E_{kin} and θ measured outside the crystal give the correct value of $k_{||}$ of the initial state inside the crystal. This information alone allows one to map the bands of two-dimensional localised states such as Shockley surface states, but also the two-dimensional band structure of an

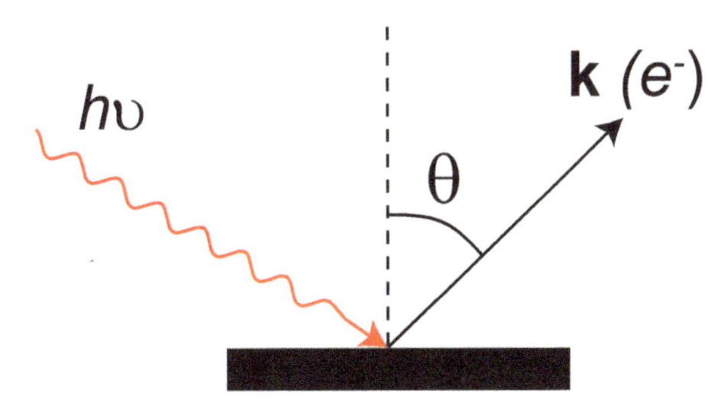

Figure 3.17. Basic geometry of an ARPES experiment showing the incident photons, energy $h\nu$, and the emitted electron (e^-) with a wavevector **k** at an angle of emission relative to the surface normal of θ.

increasing number of single-layer materials such as graphene, and of both natural and artificially 'engineered' multilayer materials in which the interlayer coupling is by weak van der Waals interactions.

Early examples of the use of ARPES to study these two-dimensional bands include investigations of the Shockley surface states that occur within a few tenths of an eV of the Fermi level at Cu(111), Ag(111) and Au(111) surfaces. Of course, these metals do not have an absolute band gap in their bulk electronic structure, but the projection of the electronic structure in the [111] direction does have a band gap around the Fermi level, and electronic states that exist in this gap are localised at the surface due to their inability to hybridise with bulk states. These states are itinerant parallel to the surface and so tend to disperse in $k_{||}$ in a free-electron-like manner. Early ARPES experiments were typically conducted using electrostatic dispersive electron energy analysers that could be moved within the UHV analysis chamber to collect the photoemission spectra in different emission directions sequentially. However, a very important instrumental development that is now exploited at all ARPES beamlines, and in very many home laboratories for a range of photoemission experiments, is the incorporation of two-dimensional detectors at the location of the original exit slits of concentric hemispherical electron analysers (CHAs, also referred to as hemispherical deflection analysers, or HDAs), allowing simultaneous detection of a range of emission angles and energies. Such a device is shown schematically in figure 3.18.

The use of this type of device allows the energy dispersion of emission from a two-dimensional band state as a function of $k_{||}$ (proportional to sin θ) to be 'imaged' directly. An example of the application of ARPES to the mapping of the Shockley surface state on Au(111) is shown in figure 3.19. Figure 3.19(a) shows individual photoelectron energy spectra recorded at different emission angles relative to the surface normal, recorded sequentially in a study reported in 1987 by Kevan and Gaylord. Figure 3.19(b) shows the two-dimensional band dispersion extracted from the shift of the peak energy as a function of angle (and thereby $k_{||}$) obtained from these spectra. Figure 3.19(c) shows the results of a more recent investigation of the same state presented in the form of a direct 'image' of a two-dimensional detector using an analyser like that shown in figure 3.18.

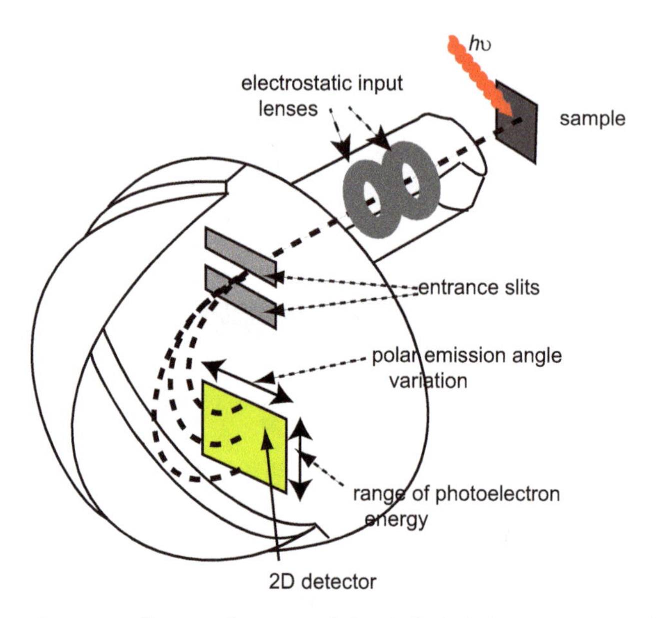

Figure 3.18. Schematic cutaway diagram of a concentric hemispherical electron energy analyser fitted with a two-dimensional detector. In this diagram the dispersion plane is vertical so short electron energy spectra appear in the vertical direction of the detector, while different emission angles are collected at different horizontal positions on the detector.

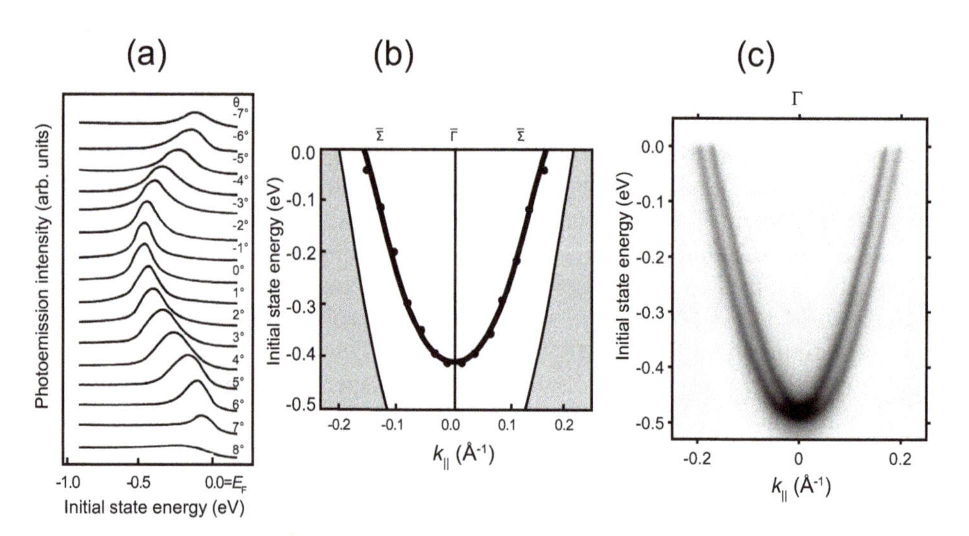

Figure 3.19. ARPES data from the Shockley surface state on the Au(111) surface. Panel (a) shows individual photoelectron energy spectra recorded at different polar emission angles from states within 1 eV of the Fermi level using a single-channel movable analyser, while panel (b) shows the E–$k_{||}$ two-dimensional band obtained from these spectra. The shaded areas are the projection of the bulk bands. Panel (c) shows the direct imaging of the same state using a multichannel analyser with two-dimensional detector. Panels (a) and (b) reprinted with permission from Kevan and Gaylord (1987), copyright (1987), by the American Physical Society. (c) Courtesy Phil King, University of St Andrews.

The improved spectral resolution of the data in figure 3.19(c) reveals a splitting of the band attributed to spin–orbit coupling; this is a manifestation of the Rashba effect and a consequence of the lack of inversion symmetry of the potential well at the surface in which the electrons are confined (LaShell *et al* 1996). More recently, ARPES two-dimensional band mapping has been applied to a wide range of 'two-dimensional materials' including true single-layer materials such as free-standing graphene monolayers, but also to a range of both naturally occurring solids and fabricated metamaterials in which the atomic layers interact only through weak van der Waals forces, the electrons therefore being strongly localised within two dimensions. Figure 3.20 shows an example of a band mapping image of the valence π-bands around the K point of the Brillouin zone of a bilayer of graphene grown on SiC(0001). Interaction between the two layers leads to a splitting of the bands, while interaction with the substrate leads to the dispersion not being linear at the Dirac point (Ohta *et al* 2006). A review by Boswick *et al* (2009) describes much of the experimental work on single-layer and multilayer graphene and the resulting understanding of the electronic structure.

While the use of concentric hemispherical dispersive analysers fitted with two-dimensional detection such as that shown in figure 3.18 is now widespread, a slightly different type of detector that is now gaining use, which also provides direct 'imaging' of two-dimensional band structures in ARPES, is the momentum

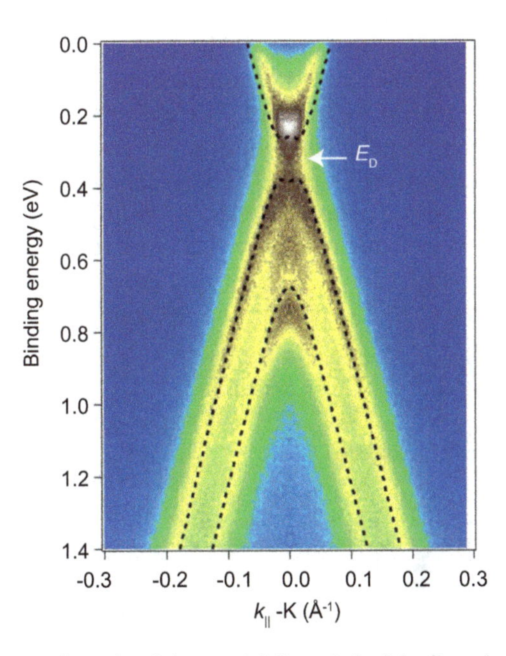

Figure 3.20. Image from a two-dimensional detector (cf. figure 3.4) of the dispersion of the π-bands of bilayer graphene around the K point of the two-dimensional Brillouin zone grown on SiC(0001), recorded at a photon energy of 100 eV. Superimposed are fitted tight-binding calculated bands. Charge transfer from the substrate leads to an offset in the energy of the Dirac point, E_D. Reprinted by permission from Springer Nature, Razado-Colambo *et al* (2018), copyright (2018).

microscope. The word 'microscope' here is the key to its mode of operation, which is a natural development of the low-energy electron microscope (LEEM) first invented by E Bauer in the 1960s; the historical context of the development of this instrument is described in a much later review (Bauer 2012). The underlying principle of the LEEM (exploiting back-scattering of electrons from a surface at energies of only a few eV) is the same as that of a transmission electron microscope (TEM) operating at electron energies of tens to hundreds of keV, and indeed is the same as that of an optical transmission microscope as described by the Abbé theory. Specifically, the Abbé theory tells us that illuminating a sample in a microscope leads to the formation of a diffraction pattern, which is the Fourier transform of the sample, but the objective optics of the microscope then generates an image that is the Fourier transform of the diffraction pattern. The quality of the image is limited by the angular aperture of the objective, which limits the number of Fourier components in the diffraction pattern that can be inverted. Figure 3.21 shows a highly simplified schematic of the components of an optical microscope, showing the diffraction and image planes.

The momentum microscope (and also the LEEM) exploits the fact that, by suitable modification of the electron optics and the location of apertures in a microscope, it is possible to display directly either the diffraction pattern or the physical image. Of course, in both the TEM and the LEEM, what are detected are the transmitted or reflected electrons resulting from the initial incident electron beam. In a photoelectron emission microscope (PEEM), operated in the same way, it is possible to obtain a physical image of the surface generated by photoemitted electrons, but in addition, when operated under conditions that would collect the diffraction pattern in a LEEM instrument, one collects the angular distribution of the photoelectrons. This angular distribution, of course, is determined by the momentum of the photoelectrons (parallel and perpendicular to the surface), leading to the name of 'momentum' microscope. One further important instrumental feature of the LEEM and PEEM instruments (described more fully in chapter 5), which also impacts of the utility of a PEEM instrument as a momentum microscope, concerns the energy of the electrons passing through the electron optics. As mentioned above,

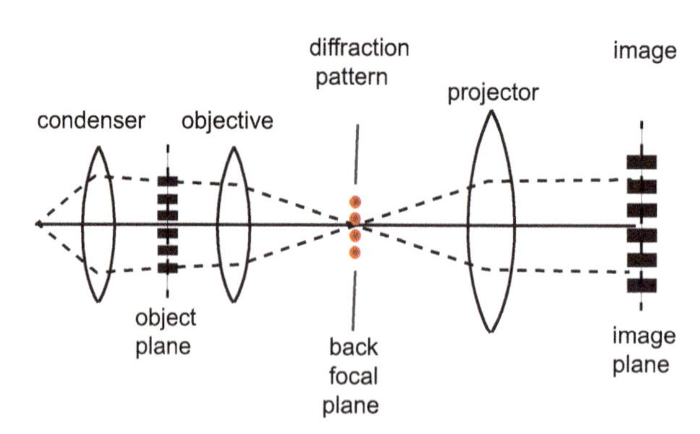

Figure 3.21. Highly simplified schematic of an optical microscope, showing the diffraction and image planes; the usual eyepiece is replaced here by a projection lens.

LEEM and PEEM instruments create images from low-energy electrons, typically no more than a few eV in LEEM but generally no more than a few tens or hundreds of eV in PEEM. The effects of space charge, as well as of stray electrostatic and magnetic fields, would severely restrict the spatial resolution that could be achieved if electrons at such low energies passed through the imaging optics. The solution of this problem is to accelerate the electrons to much higher energies (up to tens of keV) after they have left a surface and before passing through the electron optics. In a LEEM instrument, incident high-energy electrons are decelerated as they approach the surface by having the sample at a high negative voltage relative to the incident electron optics, the diffracted electrons then being reaccelerated to enter the imaging optics. In a PEEM, only the acceleration stage is required. An important consequence of this acceleration is that *all* photoemitted electrons, including those emitted at grazing emission angles, can be collected by the imaging optics. Operated in the 'diffraction' mode, to display the angular distribution of the emitted electrons, one therefore collects the full ±90° emission angular range, providing that the photoelectron energies do not exceed some limiting energy determined by the specific design of the extractor. This leads to a significantly larger solid angle of acceptance of the emitted electrons relative to the design of the instrument in figure 3.18.

A highly simplified schematic diagram of such a momentum microscope, based on the commercial KREIOS 150, manufactured by SPECS GmbH, is shown in figure 3.22. Superficially, the design is similar to that of the CHA with two-dimensional parallel detection shown in figure 3.18, but the key difference is in the electron optics before and after the electrons pass through the CHA. The momentum microscope has the accelerating extractor, but also lenses and variable apertures that allow the detector to display either the angular distribution of the photoemission or a real-space image of the surface, in both cases at a selected photoelectron energy determined by the passage through the dispersive electrostatic field between the hemispheres. The imaging capability means that it is possible to

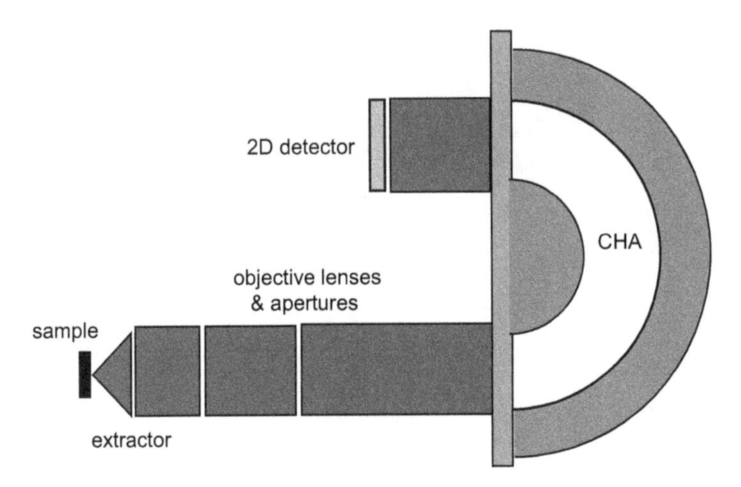

Figure 3.22. Simplified schematic diagram of a commercial momentum microscope.

select the angular distribution of electrons coming only from a small (sub-micron) area of the surface, an important feature in studying increasingly complex materials that may be only produced in small sizes or with significant spatial inhomogeneity. This general issue of different modes of obtaining photoelectron images and selected area angular distributions is discussed further in chapter 5. Much more detailed diagrams of one of these specific instruments, installed on the UVSOR synchrotron radiation source in Japan, was reported by Matsui *et al* (2020), which includes comparative information on other existing types of photoelectron microscopes installed on synchrotron radiation sources.

In addition to two-dimensional band mapping described above, ARPES measurements also allow one to determine the two-dimensional Fermi surface of two-dimensional localised states, by simply mapping the intensity of the photoemission at the Fermi level as a function of $\mathbf{k}_{||}$. The intensity is expected to peak as an occupied state disperses across the Fermi level. This approach assumes, of course, that the photoemission intensity is not strongly modulated by matrix elements effects. A particularly simple case of the application of this technique is to map the surface Fermi surface of the Shockley surface state at the Cu(111) surface, the results being shown in figure 3.23. Figure 3.23(a) shows the mapping of the experimental

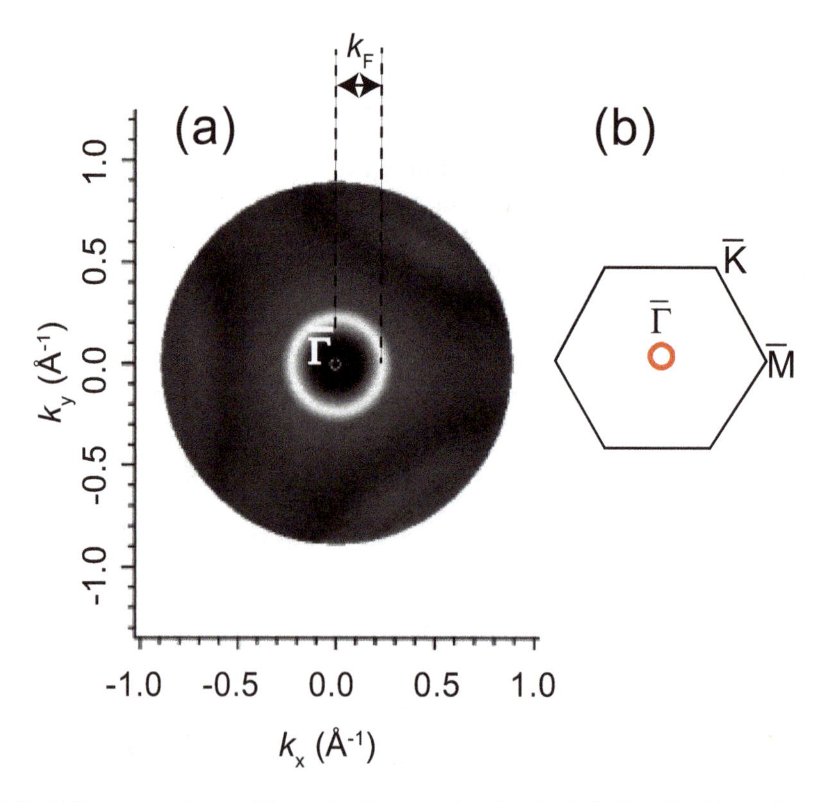

Figure 3.23. (a) Experimental map of the surface Fermi surface (bright ring) of the Shockley surface state on Cu(111). (b) The location of this Fermi surface (red) within the two-dimensional Brillouin zone. Panel (a) reprinted with permission from Baumberger *et al* (2001), copyright (2001), by the American Physical Society.

data, while figure 3.23(b) shows the Brillouin zone, with the Fermi surface of panel (a) shown in red. In this case the nearly-free-electron character of the surface state leads to its Fermi surface being circular with a radius equal to the magnitude of the Fermi wavevector. Of course, the two-dimensional Fermi surfaces of many other surface states and two-dimensional localised states are more complicated than this.

Mapping the E–$k_{||}$ relationship of two-dimensional states does not strictly require the availability of incident radiation of variable photon energy, and indeed the data of figure 3.23 were obtained using a conventional He I laboratory discharge source with a photon energy of 21.2 eV. However, the ability to move to somewhat higher energies in ARPES is essential to access a wider range of $k_{||}$ values, and indeed to ensure that a wider range of such values corresponds to the limited angular acceptance range of two-dimensional 'imaging' CHAs. In addition, however, the intensity of photoemission from two-dimensional states does depend on the incident photon energy, so the ability to vary the photon energy can be valuable in tracking the dispersion of these states.

One origin of this intensity variation, and the full impact of the ability to vary the photon energy in ARPES studies, is related to the value of $k_\perp$ of the outgoing photoelectron, especially for band mapping of 3d itinerant states. This quantity (in Å^{-1}) is also determined by the kinetic energy (in eV) and emission angle of the photoemission by

$$k_\perp = 0.5123\sqrt{(E_{\mathrm{kin}}\cos^2\theta + V_0)},$$

where V_0 is the inner potential, the difference in kinetic energy of the electron inside and outside the crystal, a quantity approximately equal to the sum of the Fermi energy and the work function of the surface, typically ~15 eV.

As remarked earlier in this section, the fact that photoemission samples only the near-surface region of a sample due to inelastic scattering means that a crystalline sample in a photoemission experiment cannot be regarded as truly three-dimensionally periodic, so while the two-dimensional periodicity parallel to the surface ensures that the components of $\mathbf{k}_{||}$ ($k_{||x}$ and $k_{||y}$) are conserved, $k_\perp$ is not strictly conserved. However, the fact that one samples several near-surface layers means that a vestige of $k_\perp$ conservation is retained. In effect, the limited sampling depth leads to a 'smearing' of this conservation law. In practice, this residual $k_\perp$ conservation allows 'bulk' band structures to be mapped with ARPES. Figure 3.24(a) shows the results of an early experiment illustrating this effect. Specifically, it shows a sequence of photoemission energy spectra recorded in normal emission from Cu(111) using a series of different photon energies. Figure 3.24(b) shows the relevant part of the band structure of Cu in the direction corresponding to normal emission from this surface. Superimposed on this figure are possible $k_\perp$-conserving transitions at two different photon energies within the range used in the experiments. Notice that the Λ_1 s–p bands show very similar behaviour to the schematic nearly-free-electron bands of figure 3.16. As the photon energy increases, 'direct' or 'vertical' $k_\perp$-conserving transitions correspond to emission from lower-energy initial states as shown by the two example photon energies. This leads to the shift of the kinetic energies of these Λ_1

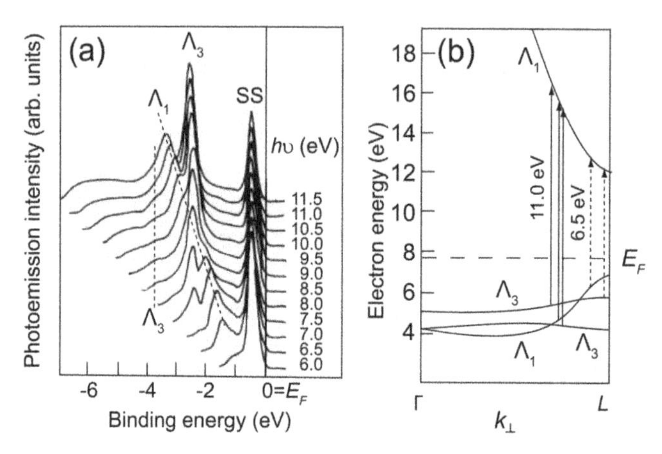

Figure 3.24. (a) ARPES spectra recorded from Cu(111) at normal emission using a series of different photon energies. (b) The relevant part of the calculated band structure, with superimposed transitions at two different photon energies. Panel (a) reprinted with permission from Knapp *et al* (1979), copyright (1979), by the American Physical Society. Panel (b) after Smith *et al* (1980).

photoemission peaks shown by the sloping dashed line in figure 3.24(a). Direct transitions are also possible from the Λ_3 d-bands, but these are relatively 'flat' (weakly dispersing), so the energy of the corresponding peak in the photoemission spectra is almost independent of photon energy. These spectra also show the peak corresponding to emission from the Shockley surface state (labelled SS) close to the Fermi level, already discussed above. The energy of this peak depends on $k_{||}$ (cf. the behaviour of the similar surface state on Au(111) in figure 3.19) but not on $k_{\perp}$, and all the spectra of figure 3.24(a) are recorded with $k_{||} = 0$ (normal emission).

This simple example highlights the necessity of variable photon energies to map the three-dimensional bands of bulk crystalline solids, and thus a crucial role of synchrotron radiation in these experiments. Notice, too, that the influence of $k_{\perp}$ conservation can also account for the photon energy dependence of the photoemission intensity from surface-localised states mentioned earlier. Specifically, the Shockley surface state on Cu(111) has a binding energy that falls in the band gap at the L point seen in figure 3.24(b). The fact that it is in this band gap means that there are no bulk states available for hybridisation, so its wavefunction must decay exponentially with distance into the crystal. This can be represented by an imaginary component, ik_i, of its wavevector, but its location at the L point means it must have a real part to its wavevector corresponding to that of the L point. This remnant of its $k_{\perp}$ value means that photoemission is most favoured when the wavevector of the photoelectron corresponds to this value, so while the $k_{||}$-conservation is the constraint that governs the possibility of photoemission from this state, its intensity is influenced by the residual $k_{\perp}$ conservation. A plot of the intensity of the surface state photoemission peak intensity as a function of $k_{\perp}$ around the value at the L point therefore provides a way to determine the value of k_i and its inverse, the wavefunction decay length; the sharper the peak, the longer the decay length. Kevan and Gaylord (1987) reported such measurements for the surface states on all three of the noble metal (111) surfaces, Cu(111), Ag(111) and Au(111), and discussed some of the complications of their interpretation.

This ability to map the bands of three-dimensional itinerant states also means that ARPES can be used to determine bulk Fermi surfaces. In this case, too, experiments on copper samples provide a simple example. The Fermi surface of Cu is rather well known, being the material chosen for the first measurements of a Fermi surface by Pippard (1957), through measurements of the variation with crystal orientation of the anomalous skin resistance. Cu was also the material chosen by at least three groups to illustrate the ability of ARPES measurements to determine the bulk Fermi surface (Avila *et al* 1995, Rotenberg *et al* 1996, Osterwalder *et al* 2000). An ideal free-electron metal is isotropic, so the Fermi surface is a sphere (of radius k_F; cf. the circular two-dimensional Fermi surface of the essentially free-electron-like Shockley surface state on Cu(111) shown in figure 3.19). However, the influence of d-band electrons leads to the Fermi surface of copper having 'necks' that link the Fermi surface in adjacent Brillouin zones (figure 3.25(a)), leading to the characteristic 'dog's bone' features in an extended-zone representation (figure 3.25(b)).

The basic principle of using ARPES to determine the (three-dimensional) Fermi surface is the same as that of mapping the Fermi surface of a two-dimensional state, namely, to map the intensity of photoemission at the Fermi level as a function of the momentum transfer. At a single photon energy ARPES provides a map of these intensities as a function of $k_{||}$, but of course at a single photon energy the value of $k_\perp$ also varies across this map, so the cut of three-dimensional k-space is a section of a spherical arc, as shown in figure 3.25(c). Changing the photon energy changes the radius of this arc, so a series of such measurements allows one to map out the

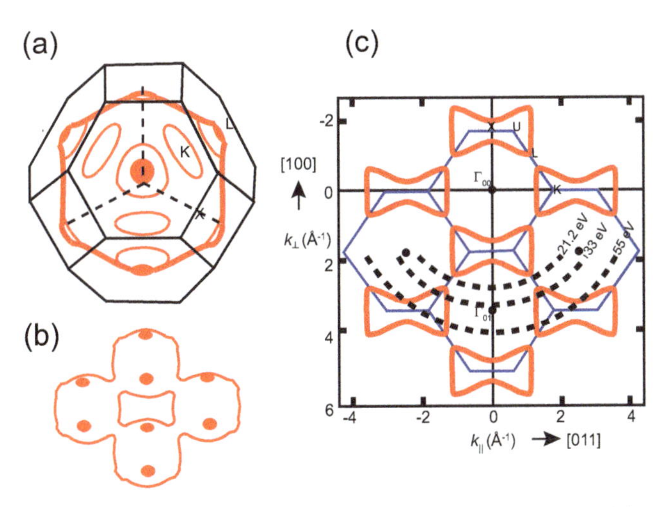

Figure 3.25. Fermi surface of Cu (red) shown in the first Brillouin zone (a) and in an extended-zone representation (b). Panel (c) shows the projection of the Fermi surface (red) onto the $(0\bar{1}1)$ plane perpendicular to the (100) surface in the [011] azimuth. The Brillouin zone's boundaries are shown in blue while the dashed lines show the sections probed by ARPES at different photon energies.

complete three-dimensional Fermi surface. Figure 3.26 shows the results obtained by Avila *et al* (1995) for the intensity of the photoemission at the Fermi level as a function of $k_{||}$ from a Cu(100) surface at two of the photon energies shown in figure 3.25(c). Also shown are the results of theoretical simulations based on tight-binding calculations, which clearly show excellent agreement with the experimental data. These experiments were actually performed using sequential measurements of photoelectron energy spectra at different emission angles using a moveable analyser, but of course complete maps of this type can now be collected in a single measurement using a two-dimensional display analyser.

While these somewhat historical examples of the use of synchrotron radiation ARPES to understand the two-dimensional and three-dimensional electronic

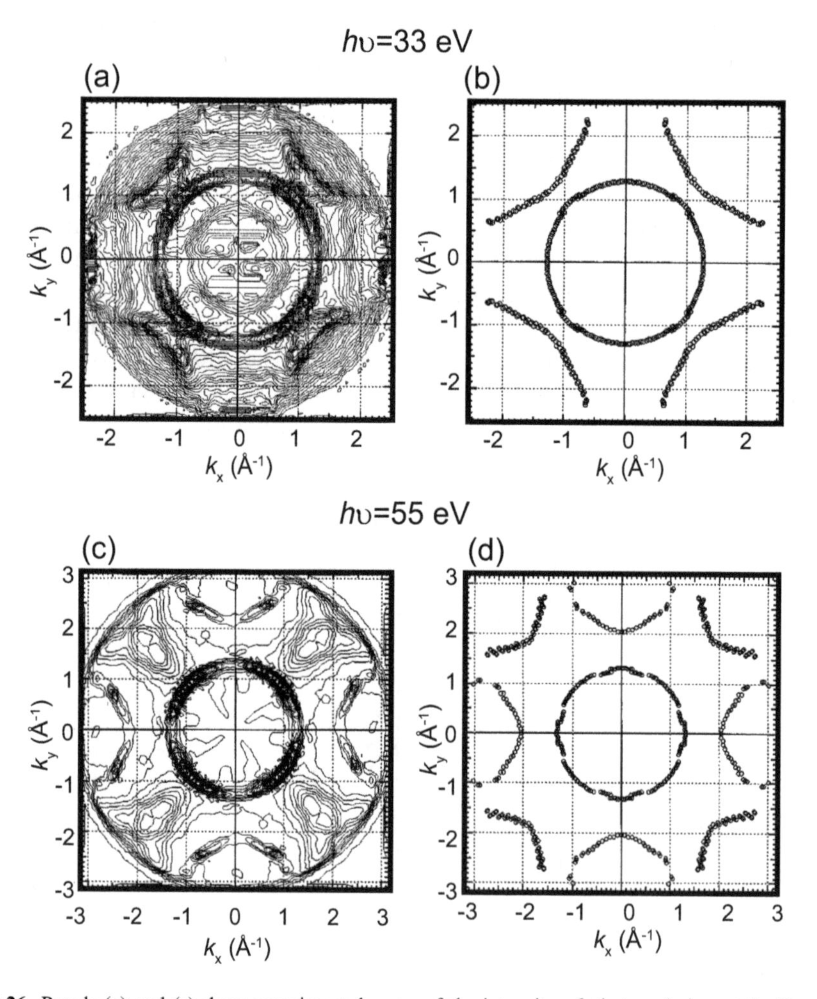

Figure 3.26. Panels (a) and (c) show experimental maps of the intensity of photoemission at the Fermi level from a Cu(100) surface as a function of $k_{||}$ at photon energies of 33 and 55 eV, respectively, and panels (b) and (d) show the results of tight-binding theoretical calculations of the results expected at these energies. Reprinted with permission from Avila *et al* (1995), copyright (1995), American Vacuum Society.

structure of the surfaces and bulk of copper illustrate the basic methodology applied to a simple system, more recent applications of this approach have addressed the electronic properties of more novel and challenging issues in condensed matter physics, such the properties of strongly correlated materials. For example, it is widely acknowledged that the application of ARPES has had a significant impact on our understanding of the so-called high-T_c (high critical temperature) superconductors, such as the cuprates, as well as a range of quantum materials (e.g., Valla *et al* 1999 and references therein; Sobota *et al* 2021). These are essentially two-dimensional materials insofar as conduction is in atomic planes that are only weakly coupled to adjacent planes, thereby being ideally suited for investigation by ARPES. For many of these systems, it is important to have high resolution in the energy or the momentum, or both. The spectral resolution is determined by a quadrature sum of the energy spread of the incident radiation, which is determined by the monochromator, and the resolution of the electron energy analyser. As discussed in chapter 2, monochromators in the VUV and soft X-ray spectral ranges typically have an approximately constant resolving power ($h\nu/\Delta h\nu$), so the best energy resolution, $\Delta h\nu$, is achieved at the lowest photon energies. On the other hand, as remarked above, low photon energies lead to low maximum values of $k_{||}$, which may prevent access to the full range of values of the first Brillouin zone. Typical optimum values of the achievable photon resolution are of the order of a few meV. The resolving power of dispersive electron energy analysers is determined by the ratio of the radius of the analyser relative to the slit width, so in this case, too, the best resolution is achieved at the lowest electron energies. However, the energy determining this resolution is the energy of the electrons inside the analyser (known as the 'pass energy'), not the energy of the electrons before entering the analyser. One can therefore achieve high electron energy resolution at higher photoelectron energies by retarding the photoelectrons as they enter the analyser electrostatically. In the case of the CHA of figure 3.18, this retardation is achieved by cylindrical electrostatic lenses to optimise the performance, also typically leading to optimum resolutions in the meV range. Notice that the intrinsic linewidth of the He I radiation traditionally used in laboratory-based ultraviolet photoemission experiments is only about 1 meV, so this is intrinsically a high-resolution source, but with no tunability of the photon energy.

One reason for wanting to perform photoemission experiments with high spectral resolution is to understand the detailed line shape of spectral peaks and the impact of many-body effects in photoemission. So far, all discussion has assumed a one-electron picture. For example, an intrinsic factor broadening peaks in energy distribution curves (EDCs) like those of figures 3.19(a) and 3.24(a) is the lifetime of the hole state left in the final state of the photoemission process. Extracting this parameter from an EDC recorded as the initial state crosses the Fermi level is complicated by the influence of the Fermi function defining the cutoff of the occupied states, and by the influence of inelastic scattering to produce a background and a low-kinetic-energy 'tail'. Specifically, omitting the effect of extrinsic backgrounds and instrumental broadening, the photoemission intensity from the valence state can be written as

$$I(\mathbf{k}, E) = |M|^2 A(\mathbf{k}, E) f(E),$$

where M is the matrix element, $M_{\mathrm{f,\,i}}^{k} = \langle \phi_{\mathrm{f}}^{k} | \mathbf{A}.\ \mathbf{p} | \phi_{\mathrm{i}}^{k} \rangle$, $\mathbf{A}$ is the vector potential of the radiation, and $\mathbf{p}$ is the momentum operator. $f(E)$ is the Fermi function, ensuring photoemission can only arise from occupied states, and $A(\mathbf{k},E)$ is the spectral function:

$$A(\mathbf{k}, E) = \frac{1}{\pi} \frac{\mathrm{Im}\Sigma(E)}{[E - E_b(\mathbf{k}) - \mathrm{Re}\,\Sigma(E)]^2 + [\mathrm{Im}\Sigma(E)]^2}.$$

$E_b(\mathbf{k})$ is the dispersion relation of the 'bare' (non-interacting one-electron) band, while $\Sigma = \mathrm{Re}\,\Sigma + i\mathrm{Im}\Sigma$ is its many-body correction, known as the self-energy. In the definition presented here any dependence of Σ on $\mathbf{k}$ is assumed to be weak, so this dependence is neglected. $\mathrm{Re}\Sigma(E)$ causes a shift in the energy and effective mass, while $\mathrm{Im}\Sigma$ gives the inverse lifetime broadening h/τ. A common consequence of the electron–phonon interaction is the creation of a 'kink' in the band dispersion near the Fermi level (as seen, for example, in the dispersion of the Cu(111) Shockley surface state at sufficiently high spectral resolution; Tamai *et al* 2013). In investigating more complex materials with ARPES, an important goal is to extract the spectral function and, specifically, the real and imaginary components of the self-energy. One mode of data presentation that can prove helpful in this process is to exploit the fact that the raw photoemission data collected using a two-dimensional detector in a CHA or a momentum microscope provides a complete E–$k_{||}$ map of the photoemission intensity, such as that shown schematically in the lower-left-hand part of figure 3.27. The most common one-dimensional type of spectrum extracted from such data (an EDC) is a cut at constant $k_{||}$ of the intensity as a function of

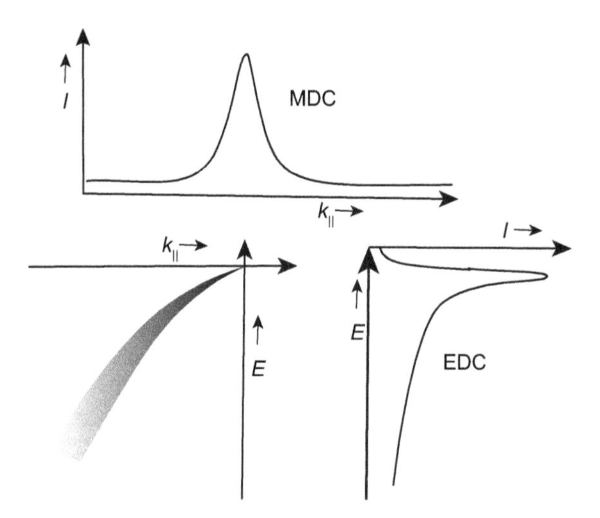

Figure 3.27. Schematic representation of photoemission data from a doped cuprate superconductor by Valla *et al* (1999), illustrating the extraction from the two-dimensional E–$k_{||}$ map (lower-left part of the figure) of an EDC (right) at the value of $k_{||}$ corresponding to the Fermi level crossing and an MDC (top) at the Fermi level.

electron energy; this is the mode of data collection of a standard moveable analyser. The lower-right-hand part of figure 3.27 shows such an EDC extracted at a value of $k_{||}$ corresponding to where the band crosses the Fermi level. The peak shape is asymmetrical due to the Fermi function cutoff and the inelastically scattered tail. By contrast, if the two-dimensional data map is cut at constant energy to produce a momentum distribution curve (MDC), as shown in the top part of figure 3.27, at an energy corresponding to the Fermi level, the peak shape is symmetrical and well-described by a Lorentzian line shape and a simple background. The full width at half maximum of the peak is $\Delta k = 2\text{Im}\Sigma(E_b)/v_0$, where v_0 is the electron velocity as determined by the band dispersion.

Figure 3.28 shows an example of the application of ARPES to determine the Fermi surface of one of the cuprate superconductors, specifically $La_{2-x}Sr_xCuO_4$ (LSCO), with four different values of the strontium concentration, x, from 0.03 to 0.22, by Razzoli *et al* (2010). For $x = 0.03$ the material is not superconducting, but for the higher values it is, with critical temperatures in the range 20–27 K; these ARPES measurements (recorded at a photon energy of 55 eV) were made below the critical temperature at 12 K. The intensity maps of figures 3.28(a–d) were constructed by reflection of the experimental data into the first Brillouin zone, using interpolation on a uniform grid and integrating the spectral weight in an energy window of 10 meV centred on the Fermi level.

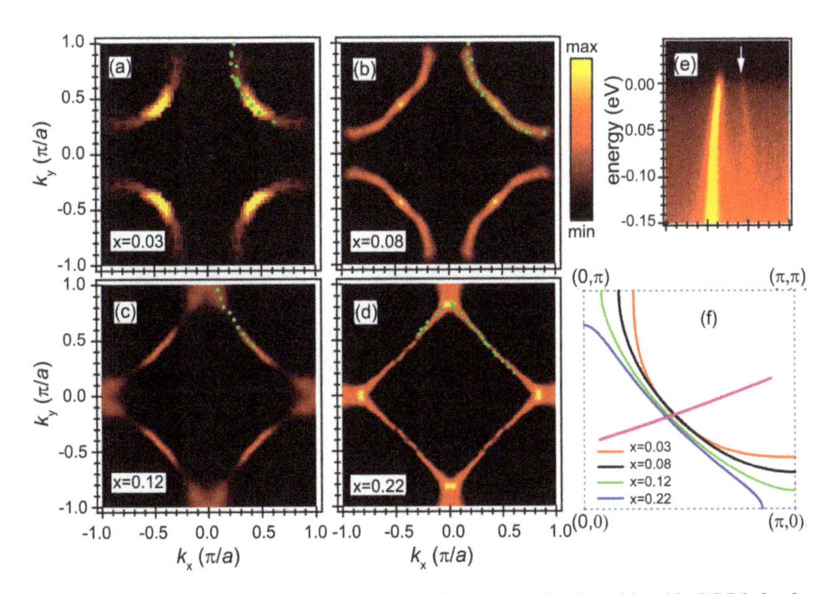

Figure 3.28. Panels (a)–(d) show spectral weight maps in $k_{||}$-space at the Fermi level in LSCO for four samples with different Sr concentrations, x. The superimposed circles are k_f values determined from peak positions in MDCs at zero binding energy. Panel (e) shows an ARPES intensity plot, acquired from the $x = 0.08$ sample, along the momentum cut indicated by the pink line in (f). Panel (f) shows the Fermi surface of each of the four samples obtained from tight-binding fits to the ARPES data. Reproduced from Razzoli *et al* (2010). © IOP Publishing Ltd. CC BY 4.0.

3.5.2 Spin-resolved ARPES

Conventional ARPES experiments determine the energy and momentum, **k**, of emitted photoelectrons, thereby allowing the energy and momentum of the initial states from which the electrons were emitted by application of the appropriate conservation laws. Additional information, however, can also be obtained (notably from magnetic materials), if one can also determine the spin state of the emitted electrons. Such experiments are certainly not new, but the loss of signal involved in determining the electron spin due to the low efficiency of spin detectors means that these experiments have been made far more viable by the enhanced photon flux at controlled polarisation delivered to the sample by modern (undulator) synchrotron radiation sources, as well as the advances in parallel detection of photoemission at different energies and emission angles, as already described.

The key requirement for these experiments is some form of spin polarimeter—a device for determining the spin of the emitted photoelectrons. Early experiments exploited the asymmetry of scattering of high-energy (~100 keV) electrons from gold atoms due to spin–orbit coupling of the electron in the potential of the gold atom (figure 3.29). In such a Mott polarimeter the spin polarisation of the incident electrons is given by

$$P = \frac{1}{S}\frac{I_A - I_B}{I_A + I_B},$$

where I_A and I_B are the detected intensities in the opposite left and right scattered channels and S is the Sherman function of the device, which is a measure of its ability to distinguish the different spins. A basis for comparing the effectiveness of different polarimeters is the figure of merit (FOM), defined as

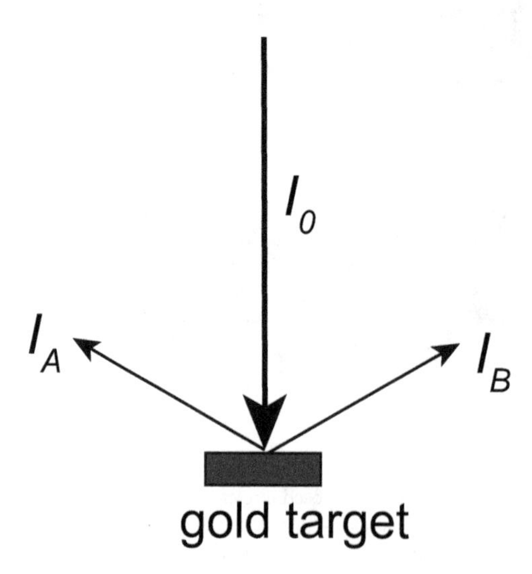

Figure 3.29. Basic geometry of a Mott polarimeter omitting all the detailed instrumental aspects such as apertures, electron lenses and current detectors.

$$\text{FOM} = S^2 \frac{I}{I_0},$$

where $I = I_A + I_B$ is the sum of the currents collected at the two detectors and I_0 is the incident beam current. The FOM of a traditional Mott detector is $\sim 10^{-4}$, while the fact that it requires the incident electrons to be accelerated to such high energies tends to ensure that it must be physically large. However, miniaturised versions of essentially the same device ('mini-Mott's and 'micro-Mott's) operating at lower incident electron energies have been found to be effective alternatives. These are more readily compatible with installation into UHV surface science chambers (including installation on moveable electron energy analysers), and similar FOM values are achievable. A simple variant of these is the low-energy diffuse scattering (LEDS) detector, which has been used in the detection of the polarisation of secondary electrons emitted from a surface in scanning electron microscopes (SEMs) in the SEMPA (SEM with polarisation analysis) technique, to provide imaging of magnetic domains in surfaces (Scheinfein *et al* 1990). In this device the polarisation detection is also based on asymmetry of scattering of electrons from a polycrystalline Au sample, but using a detector divided into four quadrants, permitting the detection of asymmetry in two orthogonal scattering planes in order to distinguish different directions of the surface magnetisation.

Particularly successful is a somewhat different polarimeter based on the polarisation of low-energy electrons ($\sim$100 eV) diffracted from a W(100) surface, another manifestation of the spin–orbit effect on scattering from atoms of high atomic number. Interest in spin-polarisation effects in low-energy electron diffraction (LEED) date back to the mid-1970s when the development of relativistic theories of LEED (Jennings 1974, Feder *et al* 1976) were followed by the first experiments to demonstrate the effect (O'Neill *et al* 1975), which was suggested to provide a superior method of surface structure determination. While this aspect of the effect never led to widespread exploitation, the idea of using it as the basis of a practical polarimeter was demonstrated shortly afterwards (Kirschner and Feder 1979) and is now an accepted alternative to the use of low-energy diffuse scattering. A further polarimeter based on low-energy ($\sim$10 eV) electron reflection from a Fe(100) sample has also been demonstrated (Tillmann *et al* 1989), the polarisation in this case arising from the magnetisation of the sample. A minor modification of this idea, using an epitaxial film of Fe(100) grown on MgO(100), has been shown to have a greatly superior FOM of $\sim 10^{-2}$ (Okuda *et al* 2008) and is exploited in a spin-polarised ARPES beamline at the Hiroshima Synchrotron Radiation Center (Okuda *et al* 2011). Johnson (1997) and Johnson and Güntherodt (2007) summarise a range of other scattering phenomena considered as a possible basis for electron polarimeters, but also present a review of spin-polarised photoemission with particular emphasis on investigations of magnetic phenomena.

Of course, much of the original interest in spin-polarised photoemission was in surface magnetism, and early experiments to investigate surface magnetism appeared to lead to the conclusion that the surfaces of ferromagnetic materials were magnetically dead. Later work indicated this was a consequence of surface contamination. An early experiment by Brookes *et al* (1990) combining spin-polarised photoemission and an undulator beamline provided clear evidence of a magnetic resonance and surface state on Fe(001), shown in figure 3.30. Notice that the surface resonance (SR) peak is quenched by adsorption of surface O.

Applications of spin-polarised photoemission to the study of magnetic materials remain an area of significant interest. Figure 3.31(a) shows a relatively recent example of spin- and angle-resolved photoelectron spectroscopy (SARPES) data recorded in normal emission from Fe(110), in this case using a laboratory He I light source with a photon energy of 21.4 eV by Dedkov *et al* (2006); these spectra are

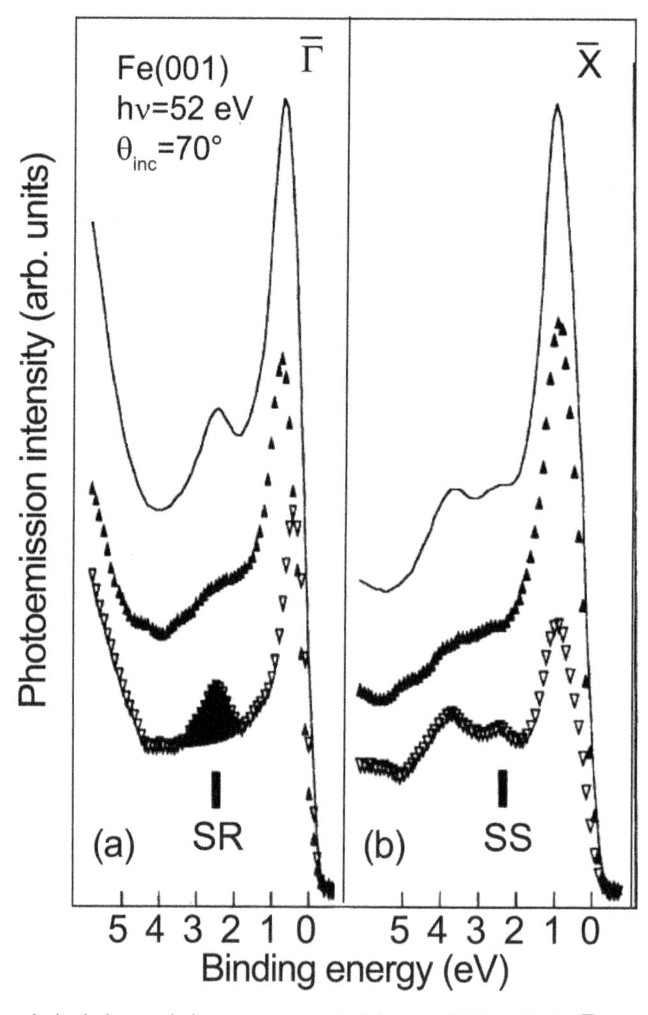

Figure 3.30. Spin-polarised photoemission spectra recorded from Fe(001) at the (a) $\bar{\Gamma}$ and (b) $\bar{X}$ points of the surface Brillouin zone. Spin-integrated data (line) are shown together with minority (▽) and majority (▲) components. SR and SS mark the minority surface resonance and surface state, respectively. The shaded region shows the attenuation of the minority surface resonance following 0.1 l of O exposure. Reprinted figure with permission from Brookes *et al* (1990), copyright (1990), by the American Physical Society.

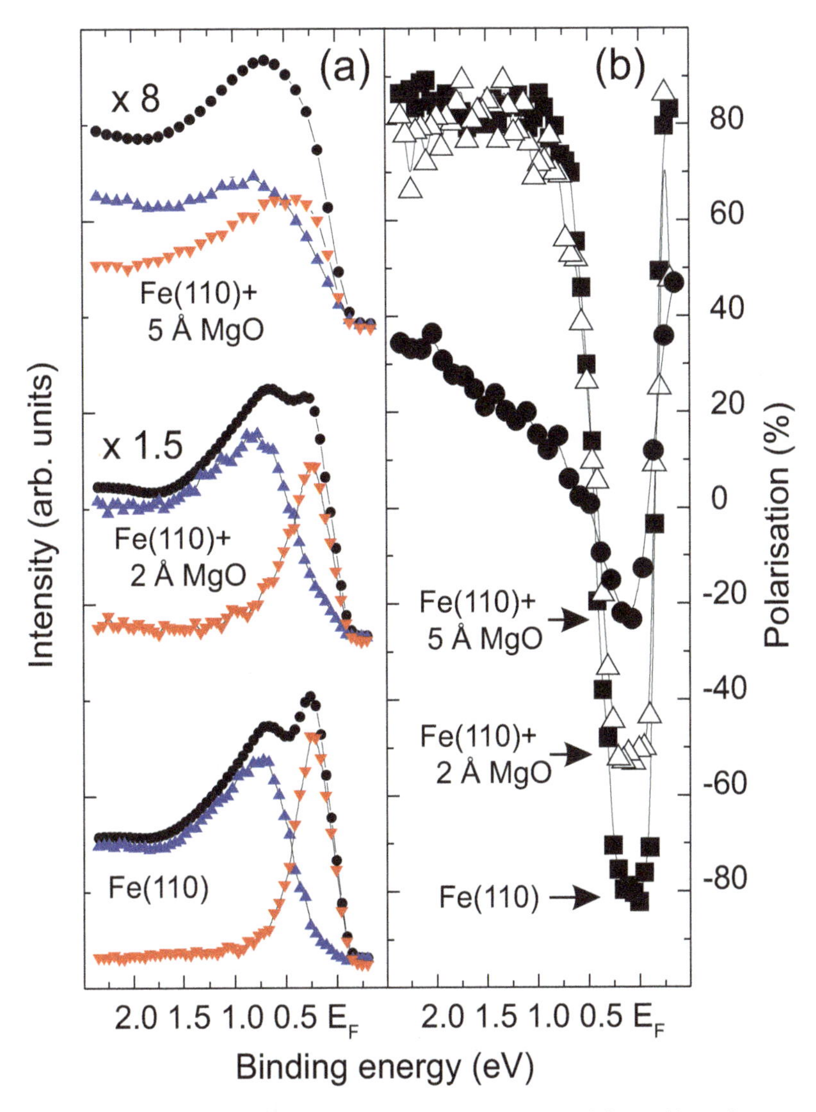

Figure 3.31. (a) Normal-emission SARPES recorded from Fe(110) and from this surface covered with epitaxial films of MgO of thickness 2 and 5 Å. Arrowheads pointing up and down show the SARPES signal for up-spin and down-spin states, respectively, while the total non-spin-resolved signal is shown as circles. Panel (b) shows the resulting variation in polarisation, the different symbols corresponding to the three different samples. Reprinted with permission from Springer Nature, Dedkov *et al* (2006), copyright (2006).

similar to those recorded much earlier by Schröder *et al* (1985) from an epitaxial Fe(110) crystal grown on gallium arsenide (GaAs).

The motivation for the investigation by Dedkov *et al* (2006) of the impact of epitaxial magnesium oxide (MgO) films on their Fe(110) sample is driven by a wish to understand the role of the interface structure in the high tunnelling magneto-resistance that can be achieved in magnetic tunnel junctions comprising two ferromagnetic electrodes separated by a thin insulating layer. The authors attribute the sharp decrease in spin polarisation at the Fermi level with increasing MgO thickness to the formation of a thin FeO layer at the interface.

A very different spin-resolved photoemission investigation of iron (Fe) in contact with MgO by Ueda and Sakuraba (2021) was performed at the very much higher photon energy of 5.95 keV in the HAXPES range. As discussed in section 3.3, a particular challenge of studying valence-band photoemission at such high energies is the very low photoionisation cross-section for these states, so introducing the further loss in signal resulting from the use of a spin detector severely exacerbates this problem. To address this difficulty the authors of this work introduced spin-selectivity *before*, rather than after, the energy selection of the standard CHA detector. This allows one to exploit the full parallel-detection capability of the CHA, rather than selecting a single channel for spin detection. This arrangement is shown schematically in figure 3.32(a). A modified sample holder in the standard HAXPES beamline holds both the sample and an Au film target from which the emitted photoelectrons are scattered into the CHA detector. Notice that, unlike a standard Mott-type polarimeter, this arrangement involves scattering only to one side, rather than having two (left, right) or four (left, right, up, down) detectors, so the spin-up and spin-down signals must be extracted by switching the direction of magnetisation of the sample, parallel and antiparallel to the incident X-ray propagation direction.

Figures 3.32(b) and 3.32(c) show, respectively, the spin-resolved valence spectra and associated polarisation from a 50 nm Fe(100) film grown epitaxially on Mg(100) and capped with a 2 nm MgO layer to protect the Fe film from oxidation. The high level of noise in these data (accumulated over a time period of 11 hr) reflects the difficulty of obtaining spin-selective photoemission data at such high photon energies. Notice that figure 3.32(b) cannot be compared directly with the ARPE spectra recorded from Fe(110) in figure 3.31(a), not only because of the different surface orientation and the fact that figure 3.32 shows angle-integrated measurement rather that the angle-resolved result of figure 3.31, but also because the very different photon energies lead to quite different relative intensities of the Fe 3d and Fe 4s orbital contributions.

A rather different application is illustrated by an investigation of the Shockley surface state on Au(111), referred to earlier, which is split (see figure 3.19(c)) due to the spin–orbit interaction and the lack of inversion symmetry in the potential well at the surface in which the surface state electrons are trapped. The results of a spin-polarised ARPES study of this surface by Hoesch *et al* (2004) demonstrated that these two states have different spin polarisations. Figure 3.33 shows some of the results of this study in the form of EDCs recorded for spin-up and spin-down electrons at a few different polar emission angles corresponding to different values of $k_{||}$ on either side of the surface normal ($k_{||} = 0$). In this experiment spin detection was achieved using a Mott polarimeter operating at an energy of 50 keV.

As remarked earlier, the use of simultaneous multichannel detection in ARPES through the use of a two-dimensional detector in a conventional CHA (figure 3.18) or a momentum microscope (figure 3.22) has hugely improved the efficiency of ARPES data collection. For spin-polarised photoemission, with an effective loss of signal by a factor determined by the FOM of the polarimeter, which may be 10^{-4} or

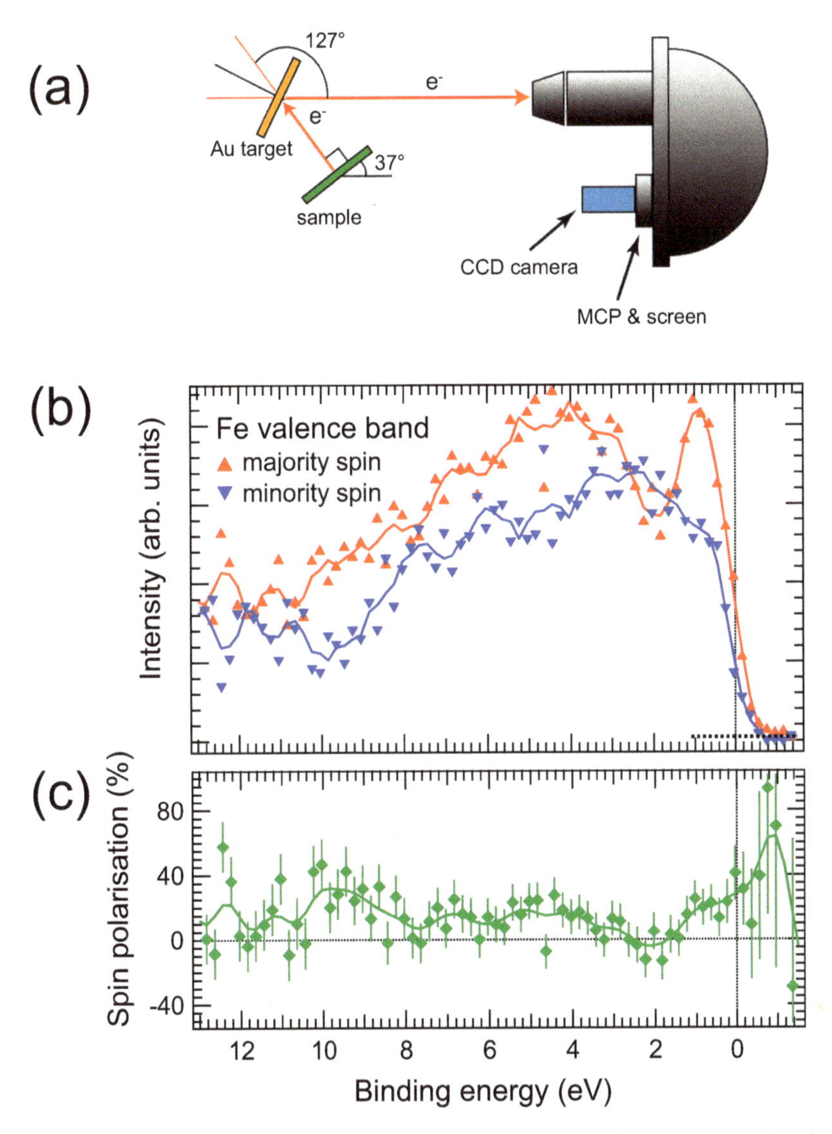

Figure 3.32. (a) Instrumental arrangement for the spin-resolved HAXPES used by Ueda and Sakuraba (2021). (b) Spin-resolved valence-band HAXPES of a 50 nm Fe(100) film sandwiched between MgO, the degree of polarisation being shown in (c). Reproduced from Ueda and Sakuraba (2021). CC BY 4.0.

less, the need for some method of parallel detection is particularly important. This HAXPES experiment addressed the problem by moving the spin selection before the CHA, maintaining the CHA's parallel detection capability but failing to exploit the k-selectivity of lower-energy ARPES studies. To achieve parallel detection in ARPES, the polarimeter must be capable of preserving the 'k-space image' that these two-dimensional detectors present to the polarimeter. This is evidently not the case if the polarimeter is based on diffuse scattering from a polycrystalline sample

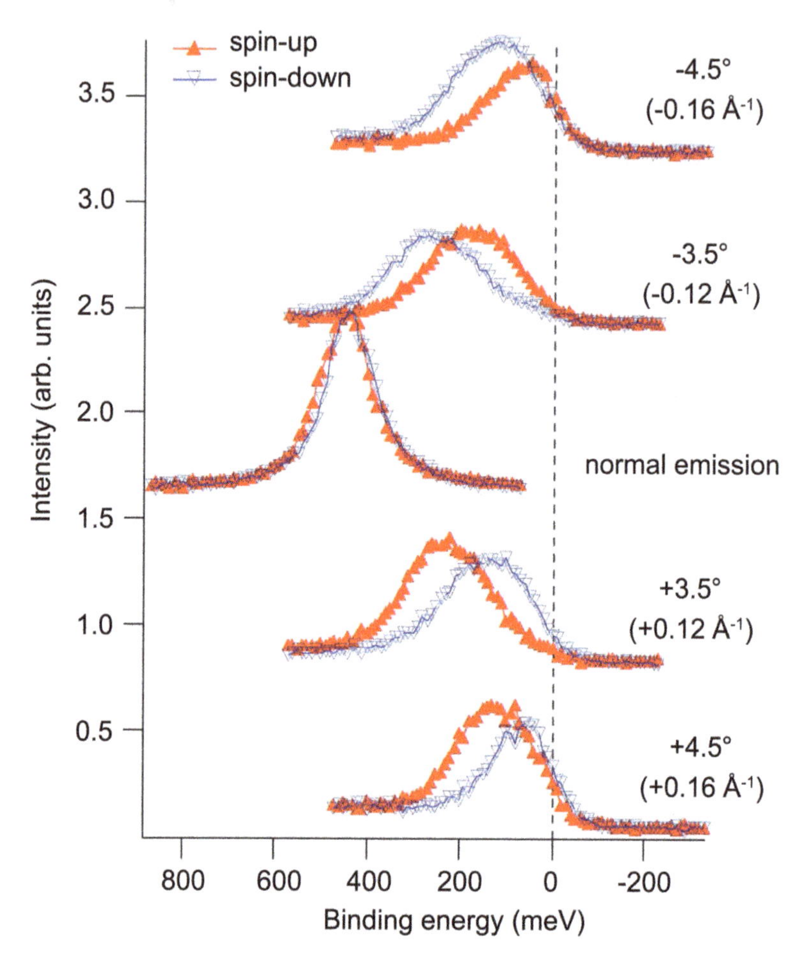

Figure 3.33. Spin-polarised EDCs recorded from the Shockley surface state on Au(111) at different emission angles, and thus different $k_{||}$ values. Reprinted figure with permission from Hoesch *et al* (2004), copyright (2004), by the American Physical Society.

(such as a Mott or LEDS detector), but can be achieved using a polarimeter based on diffraction (including specular reflection) from a single crystal scatterer. This has been demonstrated by coupling a momentum microscope to a polarimeter based on spin-polarised diffraction from an epitaxial Au(100) film (Tusche *et al* 2015); here the principle is exactly the same as the W(100) diffraction polarimeter, but the Au(100) surface is far less prone to contamination from residual gases in the UHV chamber than W(100). This type of imaging SARPES has been exploited in investigations of the topological insulators Bi_2Te_2Se (Bentmann *et al* 2021) and $TaSe_3$ (Lin *et al* 2021), the latter investigation including exploration of a strain-induced topological phase transition, although in this case the radiation source was a laser operating at a fixed photon energy of 7 eV.

Discussion of the relative merits of low-photon-energy ARPES and high-photon-energy HAXPES in studies of valence bands raises an important question, namely, is it possible to extract meaningful k-resolved band structure information (that is truly from the bulk) by performing ARPES at the high energies of HAXPES? One obvious practical difficulty arises from consideration of equation (3.1). To obtain high resolution in the measured value of $k_{||}$ when the photoelectron energy (and thus the absolute value of k) is high, one must be able to distinguish emission at very small differences in emission angle. Consider, for example, a clean low index surface of a bcc or fcc elemental solid. The distance in **k**-space to the first Brillouin zone boundary is ~1 Å^{-1}, so in a typical low-energy ARPES experiment with photoelectron kinetic energies ~50 eV (k = 3.6 Å^{-1}) this value of $k_{||}$ corresponds to an emission angle of 16°. However, if the photoelectron kinetic energy is 5 keV (k = 36 Å^{-1}), the whole Brillouin zone is captured in an emission angle range of only ±1.6°. Of course, in modern ARPES studies the samples of interest are much more complex than a simple elemental solid, so the much larger unit cell leads to even smaller values of $k_{||}$ at the first Brillouin zone boundary. A typical angular resolution of the detector ~1–2° can therefore lead to capture of at least the full $k_{||}$ range of a Brillouin zone. Of course, it is possible to reduce the acceptance angle of the analyser.

There are, however, more fundamental challenges to extracting ARPES data at specific values of $k_{||}$ at these high energies, which arise because of the dephasing effect of thermal atomic vibrations on the coherent elastic scattering of electrons in the solid. Specifically, the attenuation of the coherently diffracted intensity is given by the Debye–Waller factor, $\exp(-\Delta k^2\langle u^2\rangle)$, where $\langle u^2\rangle$ is the mean-square vibrational amplitude, measured in the direction of the scattering vector $\mathbf{\Delta k}$. The intensity that is 'lost' is spread over all other directions to produce a thermal diffuse background. In the case of an ARPES study of the E–k band structure of a crystalline solid, the quantity Δk corresponds to the reciprocal lattice vector **G** of the extended-zone representation (as shown in figure 3.16), but at high energies G is much larger than the initial-state value of k, so the appropriate value of Δk in the Debye–Waller factor is simply the final-state value of k, which in practical units (k in Å^{-1}, E in eV) can be written as $0.512E^{1/2}$, where E is the photoelectron kinetic energy.

3.5.3 ARPES of molecular orbital states

A rather different 'traditional' application of low-energy (ultraviolet) photoemission (i.e., UPS) is in the study of molecular adsorption on surfaces. The binding energies of the occupied molecular orbital states of a molecule are characteristic of the molecule, so the photoemission spectrum from these states provides a 'spectral fingerprint' of the molecule, while small shifts in the binding energy of specific orbitals provide information on the character of the molecule–substrate bonding. The combination of ARPES with the ability to vary the polarisation of the incident radiation allows one to identify the symmetry of specific orbitals.

As described in the previous section, the photoemission intensity is proportional to the modulus squared of the matrix element, $M_{\mathrm{f,i}}^{k} = \langle \phi_{\mathrm{f}}^{k}|\mathbf{A}.\,\mathbf{p}|\phi_{\mathrm{i}}^{k}\rangle$, where **A** is the vector potential of the radiation and **p** is the momentum operator. This leads to a valuable selection rule if the photoemission (the final state) is detected in a mirror plane, because in this case the final state must be either symmetric or antisymmetric relative to this plane, and if it is antisymmetric, it has zero amplitude in the mirror plane. Varying the direction of the incident polarisation vector, **A**, therefore allows one to distinguish emission from initial states of symmetric or antisymmetric character. A simple example of the use of this approach, which proved to be important in the 1970s, is in the investigation of CO adsorption on transition-metal surfaces. The spatial distributions of the three highest occupied orbitals of the CO molecule are shown schematically in figure 3.34. These are 4σ, derived mainly from hybridisation of the O and C 2s orbitals; 1π, derived from the O and C $2p_x$ and $2p_y$ orbitals; and 5σ, derived from the O and C $2p_z$ orbitals. Photoemission from a condensed layer of CO clearly shows three peaks associated with these three orbitals, but photoemission for CO chemisorbed on transition-metal surfaces shows only two peaks. The implication is that the relative binding energy of one of these states has been modified by interaction with the metal surface such that two of the states overlap in binding energy. The question is, which peak corresponds to which state?

σ-symmetry states in a diatomic molecule like CO are totally symmetric relative to the molecular axis and thus also symmetric relative to any mirror plane that contains the molecular axis. If the polarisation vector **A** is parallel to the surface ('*s*-polarisation') and perpendicular to the mirror plane (and thus also to the molecular axis), the dipole component of the matrix element is antisymmetric relative to the molecular axis. In this case the matrix element for a transition into a symmetric final state is zero, so no emission will be detected within the mirror plane

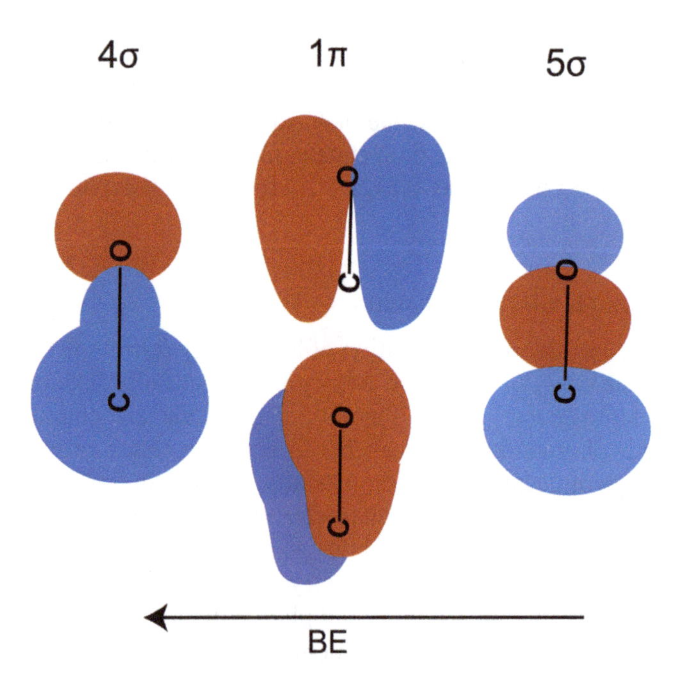

Figure 3.34. Simplified schematic diagram of the highest occupied orbitals of the CO molecule; the red and blue colouring correspond to opposite signs of the wavefunctions.

(including along the surface normal). By contrast, for emission from an initial π-state no emission will be detected in the mirror plane if the **A** vector is perpendicular to the surface ('*p*-polarisation'), thereby lying in the mirror plane. Experimental data from CO adsorbed on Ni(100) by Smith *et al* (1976) showed that in normal emission both peaks associated with the CO orbitals were observed in *p*-polarisation, but only the peak corresponding to the more shallowly bonded state was observed in *s*-polarisation. Thus, the more strongly bound state must have σ-symmetry, corresponding to the 4σ state, while the peak at lower binding energy must arise from overlapping peaks associated with the 1π and 5σ orbitals, the 5σ state being shifted down in energy due to a bonding interaction with the metal surface.

While these symmetry selection rules, combined with the availability of linearly polarised incident radiation, have proved valuable in determining adsorbed molecule orientations and orbital character, a more recent development has shown that considerably more information regarding orbital character can be obtained from ARPES studies. In particular, photoemission orbital tomography (POT), introduced by Puschnig *et al* (2009), offers the possibility of producing images of the molecular wavefunctions. As described above, the photoemission matrix element is $\langle f|\mathbf{A}.\ \mathbf{p}|i\rangle$, and exploiting the Hermitian character of the operator allows it to be applied to the final state, so if the final state is represented by a plane wave, one obtains the simple result that the angle-resolved photoemission current is proportional to the square of the product of the Fourier transform of the initial-state wavefunction and the factor **A.k**, where **k** is the photoelectron wavevector. This allows one to determine the initial-state wavefunction from an ARPES experiment. The idea of using a plane-wave representation of the final state was introduced in the early development of a theory of ARPES from adsorbed molecules in the 1970s, notably by Gadzuk (1974), but was quickly discarded as clearly incompatible with key experimental results. In particular, the **A.k** term means that the photoemission signal must be zero in any direction perpendicular to **A**. Thus, in the case of CO on Ni(100) described above, if **A** is perpendicular to the mirror plane containing the CO molecule the plane-wave approximation predicts that the photoemission signal in normal emission, or indeed in any emission angle within the mirror plane, should be zero, clearly at variance with the symmetry selection rules and the experimental data. Despite this, the plane-wave approximation in the application of the POT technique to the adsorption of (relatively) large planar π-bonded molecules has proved to be surprisingly successful when certain conditions are met.

At least in part, this can be understood in terms of key aspects of the adsorption systems to which it has been applied, but it is also helpful to describe the approach used to predict ARPES from adsorbed molecules that goes beyond the use of a plane-wave final state. In particular, Ueno *et al* (1997) used the so-called independent-atomic-centre (IAC) model of Grobman (1978), which retains the spherical-wave character of the emission treated as a coherent sum of emission from the atomic centres of the molecule, with some calculations also including the effects of the final-state multiple intramolecular scattering (Kera *et al* 2006). This final-state scattering is the basis of the structural technique of photoelectron diffraction described in section 4.4, using a core level as the initial state and exploiting the effects of final-state scattering by substrate atoms. So far, the POT technique has mostly been applied to the adsorption of relatively large and essentially planar molecules comprising only low-atomic-number elements (e.g., H, C, N, O). These weak scatterers ensure that intramolecular final-state scattering effects are unlikely to be significant, while the mismatch of intramolecular and intrasurface bond lengths ensures that the constituent molecular atoms must mostly occupy low-symmetry adsorption sites, ensuring that final-state scattering from substrate atoms will also have minimal effects on the ARPES (Bradshaw and Woodruff 2015). A further important factor is that if the detected direction of photoemission is parallel to the direction of polarisation of the incident radiation, i.e., $\mathbf{k}_{||}\mathbf{A}$, then the results of a plane-wave final state are identical to those of an exact calculation (Goldberg *et al* 1978). This condition is certainly not met by all the experimental POT studies, although the applicability of the plane-wave final state to measurements that *approximately* meet this condition is unclear.

The basic methodology of the POT technique is illustrated in figures 3.35 and 3.36 by its application to simulated data from adsorbed pentacene (figure 3.35(a)) by Offenbacher *et al* (2015). Figure 3.35(b) shows the highest occupied molecular orbital (HOMO) of pentacene in real space, while figure 3.35(c) shows the corresponding Fourier transform, its orbital in reciprocal space. The superimposed red hemisphere corresponds to a surface of constant kinetic energy, $k = \sqrt{(2m/h^2)E_{\text{kin}}}$, while figure 3.35(d) shows the absolute values of the orbital's FT corresponding to the intersection with the red hemisphere. This is therefore a simulation of an ARPES momentum map at constant photoelectron kinetic energy.

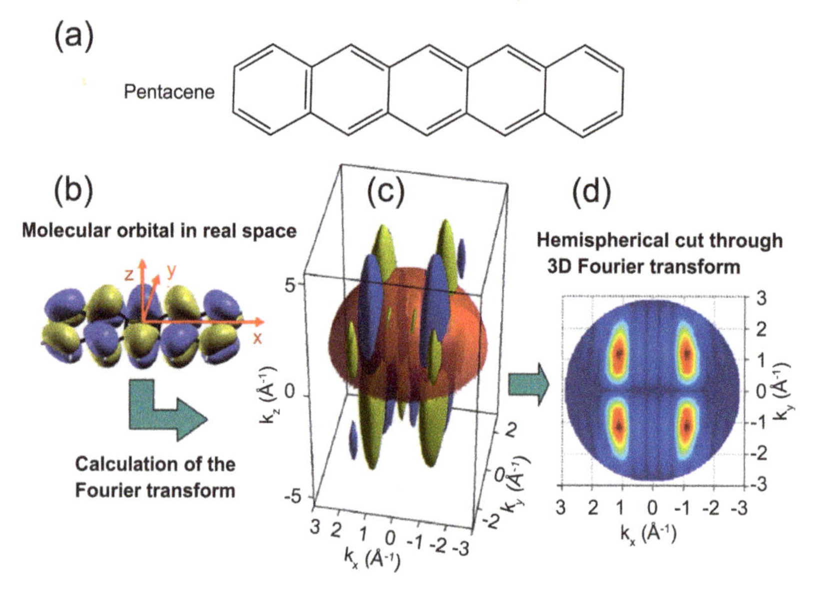

Figure 3.35. Illustration of the relationship between a calculated real-space molecular orbital (b) of pentacene (a) and its Fourier transform (c). The red hemisphere superimposed on (c) corresponds to a surface of constant photoelectron kinetic energy, while the absolute values of the FT that intersect this hemisphere are shown in (d). Reproduced from Offenbacher *et al* (2015). CC BY 4.0

Figure 3.36(a) shows the real-space DFT LUMO (lowest unoccupied molecular orbital) (10% isosurface) of pentacene, red and blue correspond to opposite signs. Figure 3.36(b) shows the Fourier transform of this orbital. Figure 3.36(d) shows the modulus squared of this transform, which corresponds to the predicted photoemission intensity map. The back transform of this with an arbitrary phase imposed leads to figure 3.36(c), which suffers from the fact that the phase information is lost in the square modulus of the wavefunction. This leads to a real-space orbital that is twice the spatial extent of the true orbital and has the incorrect phase. This phase problem is, of course, analogous to the one in X-ray diffraction and stems from the fact that one measures intensities, not amplitudes of diffracted beams and photoemission intensity. In the case of the POT technique an iterative procedure starting from random phases but applying the constraint that the true orbital size is limited to of order of the sum of the van der Waals radii of the constituent atoms, has been shown to yield meaningful phases and a good representation of the original orbital (Lüftner *et al* 2013).

The results of this phase-recovery procedure are illustrated in figure 3.37 applied to experimental photoemission data from another hydrocarbon chain molecule, sexiphenyl (figure 3.37(a)) adsorbed on Al(110). Figures 3.37(b) and (c) show the experimental ARPES momentum-dependent photoemission intensity from the highest occupied molecular orbital (HOMO) and the DFT simulation of these data, while (e) and (d) show the resulting Fourier transform (and resulting molecular orbital) and the back transform with meaningful phases after up to 250 iterations. Figures 3.37(f)–(i) show the equivalent results for the HOMO-1 state.

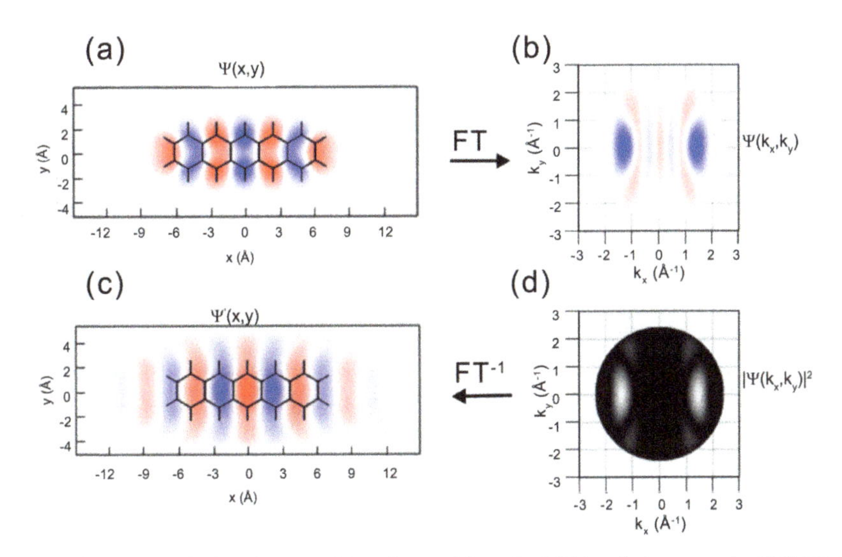

Figure 3.36. Illustration of the phase problem in applying the POT technique to the LUMO (lowest unoccupied molecular orbital) of pentacene. Panel (a) shows the real-space DFT orbital (10% isosurface) complete with the phases (red and blue shading show opposite signs) while (b) shows the Fourier transform at k = 2.5 Å^{-1}. Panel (d) shows the modulus squared of this transform, thereby simulating the experimental photoemission intensity distribution, while (c) shows the back transform with loss of meaningful phases. Reproduced from Offenbacher *et al* (2015). CC BY 4.0

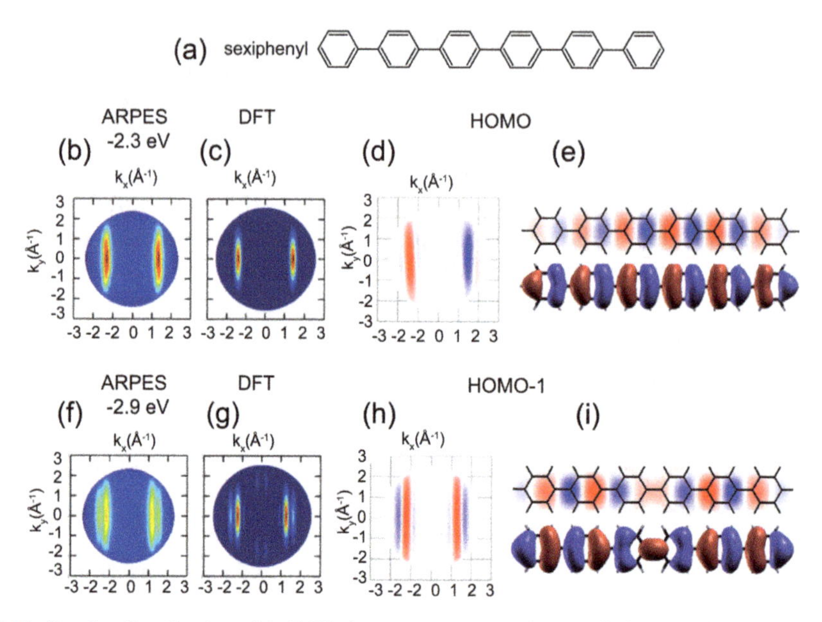

Figure 3.37. Results of application of the POT phase-recovery procedure applied to experimental ARPES data from the two highest-energy occupied molecular orbital states of sexiphenyl adsorbed on Al(110) as described in the text. Reproduced from Offenbacher *et al* (2015). CC BY 4.0

Notice that a key requirement for the application of this technique is to collect the ARPES data over a wide range of k_x and k_y—ideally over the full 2π steradians of emission. In early experiments developing this method a novel toroidal dispersive electron energy analyser (Broekman *et al* 2005) was used, although a range of more modern detectors functioning as momentum microscopes are well-suited to this task.

The particular success of the POT technique to these π-conjugated organic molecules can be understood as a consequence of the fact that the molecular orbitals are derived only from $2p_z$ orbitals on identical C atoms, and under these circumstances the plane-wave approximation is equivalent to the IAC model. Furthermore, as remarked above, the effect of final-state scattering for these large molecules comprising low-atomic-number atoms is expected to have little influence on the ARPES. These factors do not apply, however, to σ-orbitals deriving from C 2s, $2p_x$ and $2p_y$ orbitals, but a recent application of POT to σ-orbitals in 10,10-dibromo-9,9-bianthracene (DBBA) adsorbed on Cu(110) by Haags *et al* (2022) appears to have been successful despite a lack of clear theoretical understanding of why this should be the case.

3.6 X-ray absorption spectroscopy

Photoemission is essentially a probe of occupied electronic states. While ARPES studies of valence-band states in periodic structures do involve transitions to specific unoccupied final states, photoemission is not generally regarded as a probe of unoccupied states. By contrast, X-ray absorption spectroscopy (XAS) is a probe of these electronic states but is also a probe of local structure. In particular, if the energy of the incident photons exceeds the threshold energy for photoionisation by more than a few tens of eV (figure 3.38), the energy-dependent modulations observed in the photoionisation cross-section are referred to as extended X-ray absorption fine structure (EXAFS) and are associated with the coherent interference of elastic back-scattering of the photoelectrons onto the emitting atom. As described in section 4.3, EXAFS is a structural technique rather than a probe of electronic structure.

Using incident photons of energy much closer to the photoionisation threshold (typically within less than ~50 eV as shown in figure 3.38), however, the resulting XANES (X-ray absorption near edge structure) spectra are a probe of the unoccupied electronic structure, but are also strongly influenced by local geometric structure. Theoretical simulations of XANES can be achieved by calculations of the multiple scattering of electrons released in the photoionisation, so they are influenced by the local geometrical structure, but insofar as the electronic structure

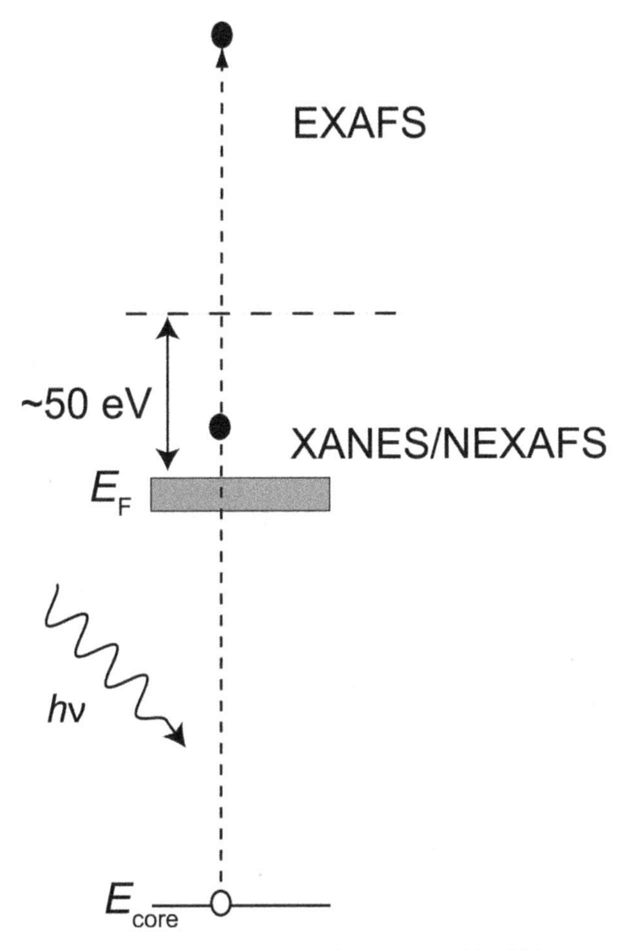

Figure 3.38. Energy-level diagram showing the XANES/NEXAFS and EXAFS ranges of X-ray absorption spectroscopy (XAS).

can also be derived from similar multiple scattering calculations, the geometrical and electronic structure are intimately related. Notice that these near-edge absorption fine-structure spectra are also frequently referred to as NEXAFS; NEXAFS and XANES are two alternative names for the same phenomenon. The NEXAFS name was introduced to describe studies of molecular adsorbates and solids, the dominant spectral features being associated with intramolecular scattering resonances; this technique is described in section 4.3 in the context of the structural information on molecular orientation that can be extracted. However, some researchers also use the NEXAFS name to describe XANES from inorganic systems, so the terminology is not consistent.

The most obvious manifestation of the sensitivity of XANES to electronic structure is in changes in the threshold energy associated with different chemical states of the photo-absorbing atom. The effect is qualitatively similar to 'chemical shifts' in core-level photoelectron binding energies, although the final states of photoemission (a core hole and an electron in the kinetic continuum) and threshold photoabsorption (a core hole and an electron in an otherwise unoccupied state just above the Fermi level) differ. Figure 3.39, which shows the results of some early S K-edge XANES measurements of S-containing petroleum asphaltenes and model compounds, illustrates both the large shifts (up to ~10 eV) in the threshold energy that can occur in different chemical states and also significant differences in the XANES structure in the 10–20 eV range above the edge.

While the quantity measured in an XAS experiment is absorbance, which can be determined by transmission through a thin sample, an alternative way to determine this quantity is to measure not the creation of the core holes but their decay. In particular, a core hole is refilled by a more weakly bound electron, the energy release being in the form of either fluorescent X-rays or Auger electrons. Auger electron detection ensures that the signal originates from the near-surface region due to inelastic scattering, so this provides a method of surface-specific XAS measurements; X-ray fluorescence samples deeper into the sample and is therefore more bulk sensitive, but may also be advantageous for studying buried interfaces and surfaces under near-ambient gas-phase conditions. An alternative mode of detection is so-called total electron yield (TEY), detecting the full energy range of emitted (secondary and inelastically scattered) electrons, or a partial yield in which the lowest-energy electrons are suppressed. These modes of measurement do not require a dispersive energy-selective electron analyser and so can accept a large angular range of emission and thus a much larger total signal, although they are less surface specific. Indeed, while TEY detection can involve collecting all the emitted electrons, as implied by its name, it can also be achieved by measuring of the sample current, thereby measuring the total *loss* of electrons. The Auger detection mode leads to a weaker signal but better signal to background; total or partial yield provide much larger signals but worse signal to background. Which method gives the best signal-to-noise ratio in the XAS modulations can be a matter of trial and error.

XANES spectra have been widely used as a spectral fingerprint of different local chemical states and structure, although quantitative modelling using multiple scattering calculations plays an increasingly important role. A particular trend is

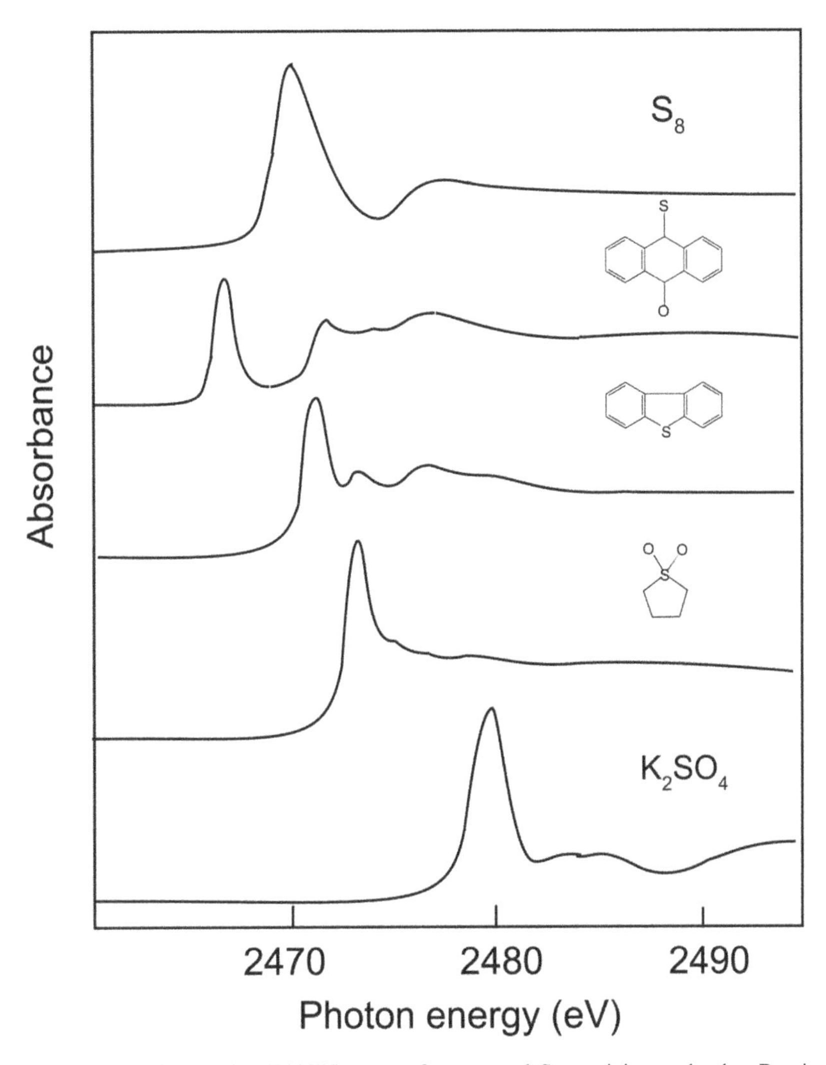

Figure 3.39. A subset of S K-edge XANES spectra from several S-containing molecules. Reprinted with permission from George and Gorbaty (1989). Copyright (1989) American Chemical Society.

in the identification and elucidation of nanostructures at surfaces including heterogeneous catalysts. Of course, heterogeneous catalysts are most commonly in the form of small particles, which provide the highest (active) surface area for a given mass, and so have long been the subject of surface science studies; nanometre dimensions, which modify the electronic and geometrical structure, can also determine their activity.

Figure 3.40 shows an example taken from an investigation of Cu K-edge XANES from Cu clusters using two alternative computational modelling codes, FEFF (Rehr *et al* 2010) and FDMNES (Bunău and Joly 2009). Notice that the XANES spectra from the smallest clusters are relatively featureless compared to those from large clusters and bulk metal.

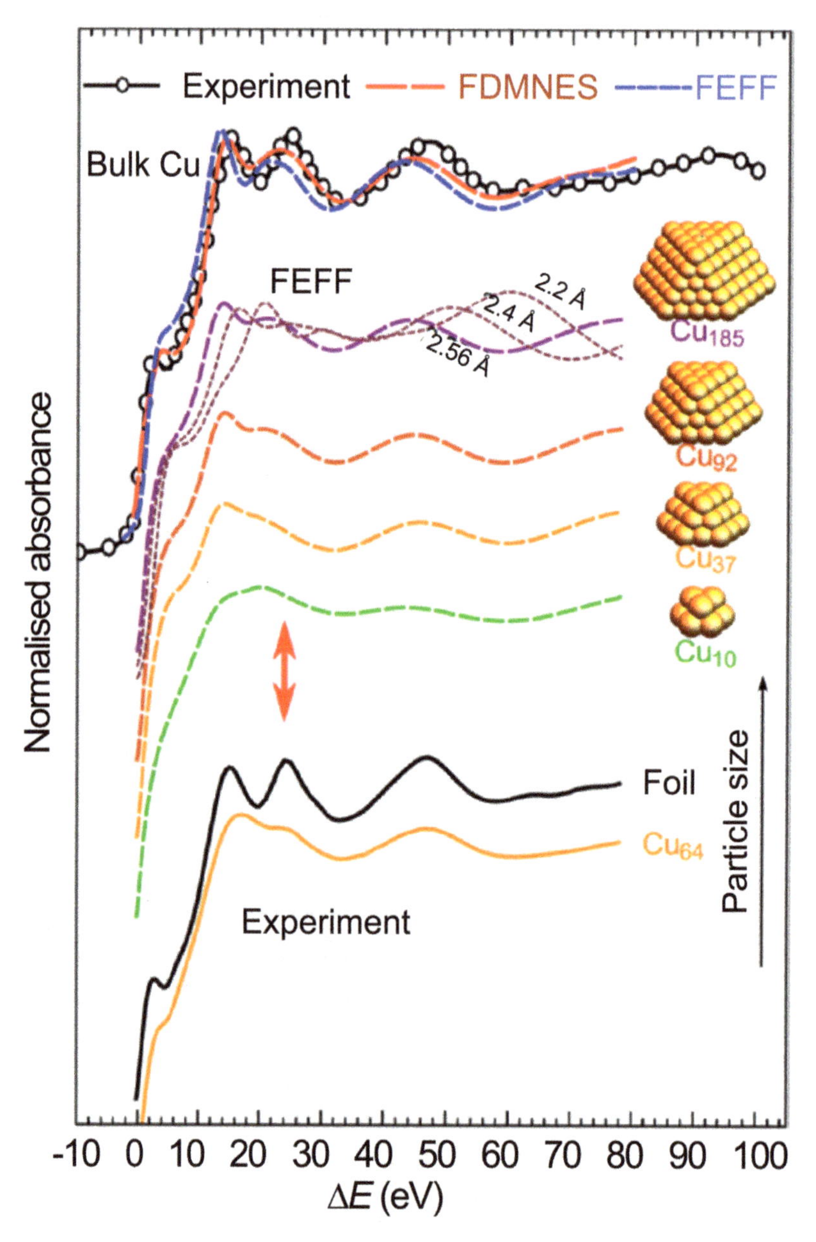

Figure 3.40. Experimental Cu K-edge XANES and theoretical spectra calculated using two alternative computer codes for bulk Cu and Cu clusters of different sizes. For the Cu_{185} cluster the results of FEFF calculations are shown for three different interatomic distances. The red arrows highlight the XANES feature most sensitive to particle size. Reprinted with permission from Timoshenko *et al* (2018). Copyright (2018) American Chemical Society.

An example of an application of Cu K-edge XANES to study the behaviour of a heterogeneous catalyst is shown in figure 3.41. Specifically, this figure shows experimental Cu K-edge XANES (recorded in fluorescence mode) from a small-pore Cu-SSZ-13 zeolite catalyst of interest for the selective catalytic reduction of ammonia. Under reaction conditions Giordanino *et al* (2014) used vibrational (infrared) spectroscopy to monitor the adsorbed ammonia and XANES to cast light on the state of the Cu in the catalyst. After activation of the catalyst the XANES is dominated by the threshold peak B assigned to the dipolar 1s→4p transition of Cu^+. The pre-edge peak A is assigned to the 1s→3d transition in Cu^{2+}; the reasons why this dipole-forbidden transition is observed are discussed, for example, by Sano *et al* (1992). The attenuation of peak A and the rise of peak B with increased ammonia exposure is interpreted as clear evidence for the reduction of Cu^{2+} to Cu^+. DFT calculations of geometry optimisation led to the two local structures shown in

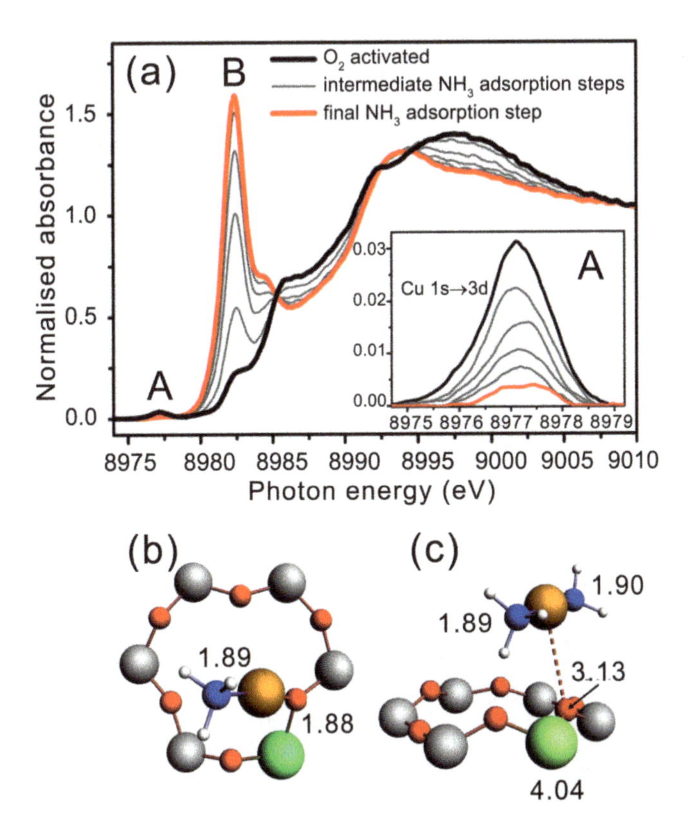

Figure 3.41. (a) *In situ* Cu K-edge XANES from the Cu-SSZ-13 catalyst after O_2 activation and during the interaction with a flow of 1300 ppm NH_3 in He at 120 °C. The inset show the pre-edge feature A magnified after background subtraction. Also shown are the local environment of Cu found in DFT calculations with (b) one and (c) two ammonia molecules, respectively (only the six-ring atoms are shown). Colour code: Cu, orange; Al, green; Si, grey; O, red; N, blue; H, white. Distances between Cu and neighbouring atoms are shown in ångström units. Reprinted with permission from Giordanino *et al* (2014). Copyright (2014) American Chemical Society.

figures 3.41(b) and (c), corresponding to one and two NH_3 molecules bonded to a single Cu site, respectively. Both structures were found to be consistent with the XANES data (and also with the results of X-ray emission measurements reported by these authors).

A rather different example of the application of XANES is in the *operando* study of a buried interface, namely, the solid electrolyte interphase (SEI) that forms on Li-ion battery anodes (Swallow *et al* 2022). This type of battery is the current dominant technology for powering portable electronic devices, but to achieve their more widespread use in electric vehicles requires improvements in their energy density and cycle lifetimes. One possible direction to achieve this goal is to replace the graphite of the standard anodes by silicon ($Li_{15}Si_4$ after lithiation). Figure 3.42 shows a schematic diagram of the electrochemical cell used in this investigation, together with graphical representations of the time dependence of the incident X-rays and the resulting current in the cell. The cell has a suspended ~100 nm-thick Si_3N_4 membrane X-ray-transparent window with an amorphous Si working electrode (WE) and a Li counter electrode (CE). The electrolyte is LP30 (lithium hexafluorophosphate, $LiPF_6$, solution in ethylene carbonate and dimethyl carbonate). The XANES measurements were made with soft X-rays around the O, F and Si K-edges in the TEY mode. When the incident photon energy exceeds the absorption edge the resulting core-level photoionisation leads to the emission of Auger electrons that lose energy in a cascade of inelastic scattering events, leading to a broad spectrum of inelastically scattered and secondary electrons. As remarked above, these electrons emitted from the surface can be measured directly in a UHV solid–vacuum interface experiment, but the loss of these electrons can also be monitored by simply measuring the sample current, providing an alternative mode of TEY detection that does not require UHV collection of emitted electrons. A complication in the case of this study of an electrochemical cell is that there is also a Faradaic current,

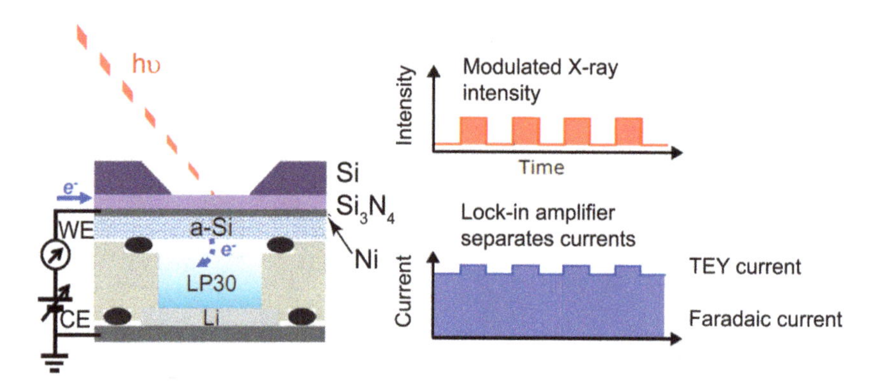

Figure 3.42. In the left is a schematic diagram of the electrochemical cell used by Swallow *et al* (2022), showing the Si_3N_4 X-ray window and the amorphous Si and Li working electrode (WE) and counter electrode (CE), respectively. On the right is shown a simple representation of the time dependence of the chopped incident X-ray flux and the resulting current that is a sum of the modulated TEY current and the unmodulated Faradaic current. Reproduced from Swallow *et al* (2022). CC BY 4.0.

but by chopping the incident X-ray flux using a motorised rotating slotted disk this Faradaic current could be separated from the modulated TEY current using a lock-in amplifier. Notice that, even for these soft X-rays, the penetration depth is around a micron, so the degree of surface specificity is determined by the average escape depth of the TEY electrons, which is around a nanometre. Complementary bulk XANES data could be collected using fluorescent yield detection.

Figure 3.43 shows O K-edge and F K-edge *operando* XANES spectra recorded at a series of different fixed potential holds. The features I to IV in the O K-edge XANES are attributed to the electrolyte solvents ethylene carbonate and dimethyl carbonate. At lower potentials these electrolyte features are no longer visible and the line shape above 534 eV closely resembles that of SiO_2 or possibly oxidised Si_3N_4. The weak I* peak that grows at the lowest voltages is attributed to the O 1s to π^* transition of carbonyl (C=O) groups associated with the growth of organic components of the SEI. In the F K-edge XANES the features I to III are attributed to PF_6^- ions, while at lower voltages below 0.6 V the spectral shape is attributable to LiF. A much more detailed discussion of the interpretation of these data and the implied growth behaviour of the SEI, aided by DFT calculations, is presented by the authors of this study. A key general conclusion is that TEY XANES, even from low-atomic-number elements using soft X-rays, can provide local structural information on 'buried' interfaces.

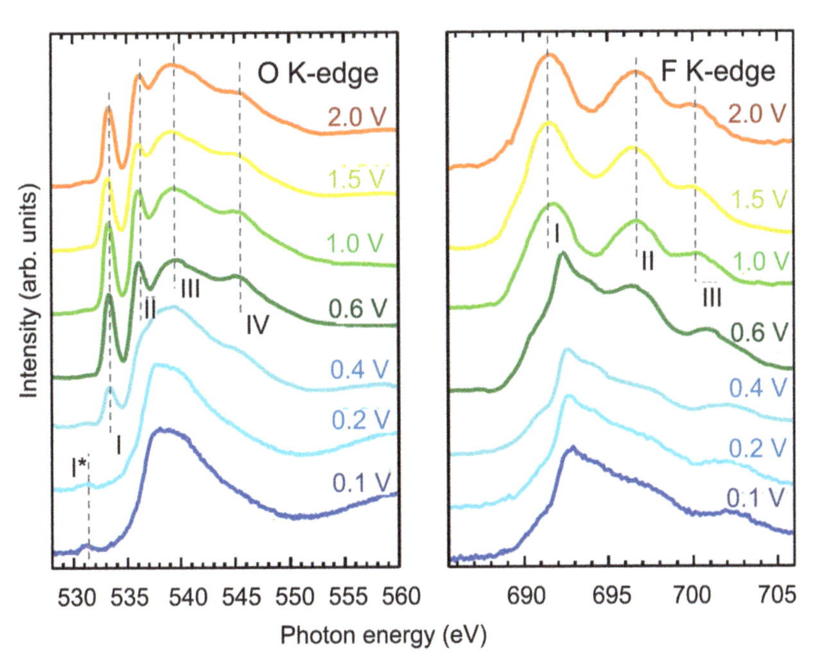

Figure 3.43. O K-edge and F K-edge *operando* XANES spectra recorded from the model LiB cell of figure 3.42. Reproduced from Swallow *et al* (2022). CC by 4.0.

3.7 Dichroism

3.7.1 Circular dichroism in the angular dependence of photoemission

While the design of undulators to provide circularly polarised radiation was described in section 2.3, the experimental techniques described so far have relied only on linearly polarised radiation. There are, however, a number of significant phenomena in surface science that rely on the use of circularly polarised radiation to measure the general phenomenon of circular dichroism (a difference in behaviour for left- and right-circularly polarised radiation) that arise from interaction of the radiation with a chiral system, i.e., a system for which its mirror image cannot be overlaid on the original. The simplest type of experiment to reveal this effect is a transmission/absorption measurement, and indeed transmission of visible radiation through an 'optically active' crystal with a chiral structure, such as quartz, has been known to lead to rotation of the plane of polarisation of linearly polarised light for more than two centuries. More recently this effect, using long-wavelength synchrotron radiation has been used to investigate folding of protein molecules. In surface science this dichroism is manifest in both photoemission and photoabsorption, with the dichroism in XANES in applied magnetic fields being particularly important.

In photoemission the effect is manifest in circular dichroism in the angular distribution (CDAD) and while this is a characteristic of photoemission from a chiral molecule, it can also arise from an achiral sample measured in a chiral experimental geometry. Specifically, the effect can be detected in photoemission from an oriented diatomic molecule if the direction of the incident radiation, the molecular axis and the direction of the outgoing photoelectrons are not coplanar; in this case, the experiment has a 'handedness' that permits the observation of CDAD. The phenomenon occurs already in the pure electric dipole approximation, but has been shown not to occur in the plane-wave approximation (Dubs *et al* 1985). Unlike photoelectron spin-polarisation effects, CDAD does not involve the electron spin.

While a chiral experimental geometry allows CDAD to be detected, of potentially greater interest are CDAD experiments from adsorbed chiral molecules in an achiral experimental geometry, with the objective of answering the question as to whether CDAD is a useful technique for identifying the chirality of adsorbed molecules. The results of a study of the adsorption of alanine, the simplest chiral amino acid, on Cu (110), provide a partial answer to this question. Figure 3.44 shows a schematic of the configuration of the intact molecule with the carbon atom at the chiral centre marked as C*. Interaction with the Cu surface leads to deprotonation of the acid hydrogen, so the molecular adsorbate is actually alaninate. Also shown is a C 1s photoemission spectrum recorded at a photon energy of 310 eV, showing that emissions from the three inequivalent C atoms are clearly resolved.

The experimental geometry for this experiment was such that the incident photons, the detector direction and the surface normal were coplanar. Setting this plane to coincide with a mirror plane of the surface (<110> or <001>) corresponds to an achiral experimental geometry, so in the absence of the adsorbed alalinate there should be no CDAD signal. However, adsorption of a chiral molecule should then produce a chiral surface, thereby leading to a CDAD signal. In addition, of

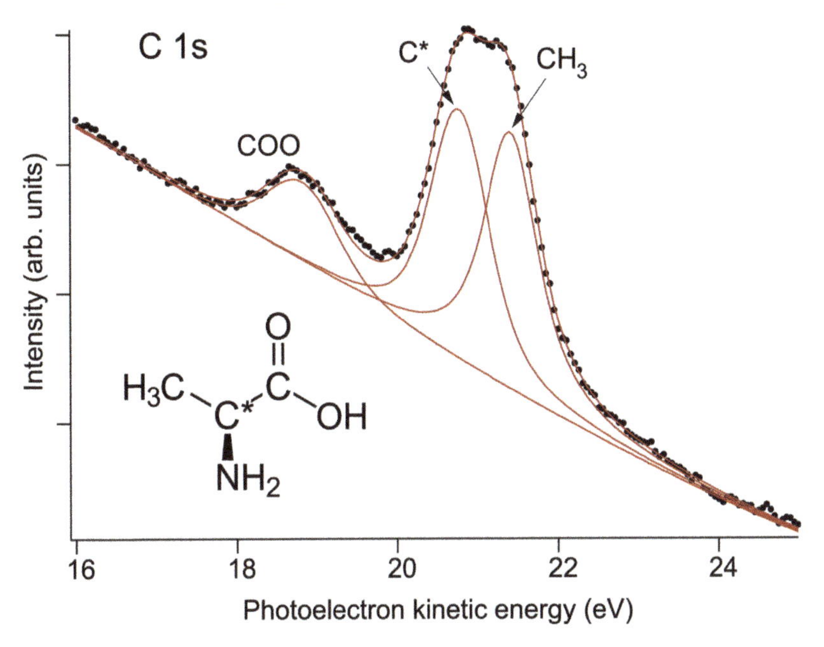

Figure 3.44. Schematic of the intact alanine molecule and C 1s spectrum from the deprotonated alaninate adsorbed on Cu(110). Reprinted with permission from Polcik *et al* (2004), copyright (2004), by the American Physical Society.

course, rotating the crystal such that the incident and detection plane no longer coincides with the crystal mirror plane will introduce a CDAD signal due to the chiral experimental geometry. Figure 3.45 shows the results of this experiment, recorded from the two opposite enantiomers (mirror images of each other) of alanine, D-alanine and L-alanine, the magnitude of the CDAD for each chemically resolved C 1s signal being recorded at different azimuthal rotations of the Cu(110) surface between the two mirror planes. Individual experimental data points are shown, joined by a bold line, while a thin line shows theoretically predicted values. Large CDAD signals of order 20% or more for the C* emissions are seen at azimuthal orientations between the mirror planes, but these are of the same sign for the two opposite enantiomers and are therefore clearly due to the chiral experimental geometry under these conditions and not primarily due to the molecular chirality. However, some key differences in the CDAD in the <001> azimuth that are equal and opposite for the two enantiomers, but typically at the level of only ~5%, are consistent with the influence of the chiral molecules. These effects are seen for all the C 1s signals, not only for that from the C* chiral centre. This may be seen to be a consequence of the fact that the adsorbed chiral molecule renders the whole surface chiral. In the gas phase one may expect the effect to only occur for the C* emission, but on a surface this is not the case.

The overall conclusion of this study is therefore that weak CDAD due to the presence of a chiral molecule on the surface can be detected, although care is required to eliminate CDAD due to chiral experimental geometry. CDAD due to an

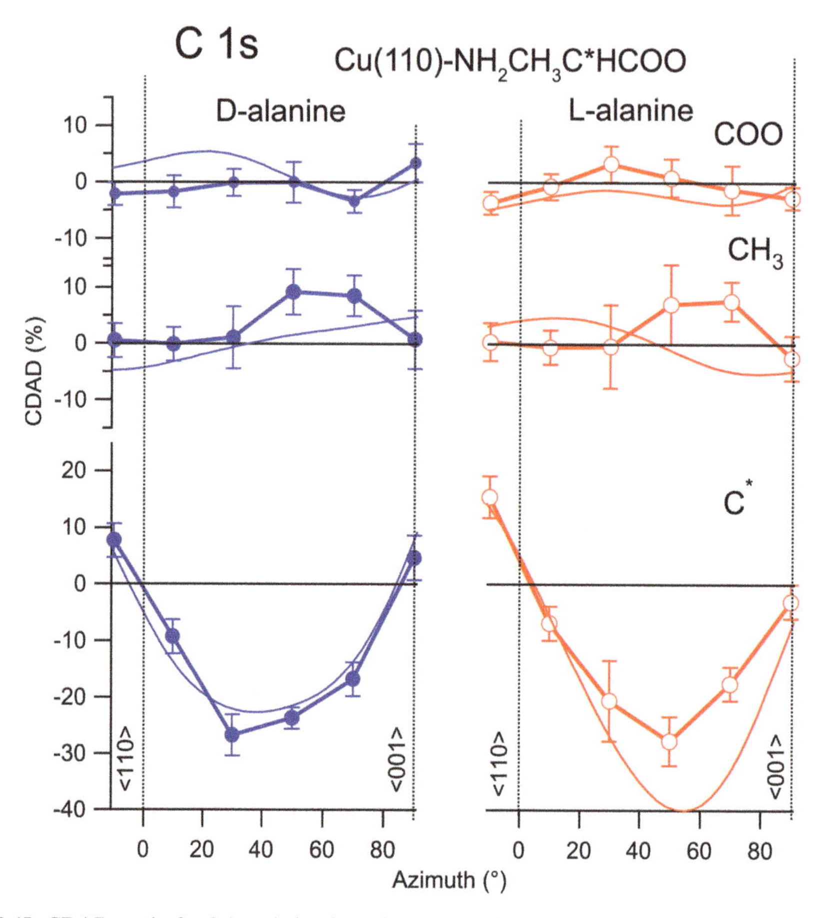

Figure 3.45. CDAD results for C 1s emission from the two enantiomers of alaninate adsorbed on Cu(110) as a function of azimuthal plane relative to the <110> and <001> mirror planes. Bold lines join individual experimental data points while the thin lines show theoretically predicted behaviour. Reprinted with permission from Polcik *et al* (2004), copyright (2004), by the American Physical Society.

adsorbed chiral molecule has also been demonstrated for 2,3-butanediol on Si(100) and tartaric acid on Cu(110) by Kim *et al* (2007).

3.7.2 X-ray magnetic circular dichroism

Measurement of circular dichroism in XANES of magnetic materials is an important technique providing detailed quantitative information about local atom-specific magnetic moments, both in bulk materials and at surfaces. X-ray magnetic circular dichroism (XMCD) can most simply be understood as a two-step process. The 2p core states of a 3d metal are spin–orbit split into $j = 1/2$ and $j = 3/2$ states at the L_2 and L_3 edges, corresponding to spin and orbital moments coupled antiparallel, and parallel, respectively. In the first step emission with the radiation helicity parallel to the 2p orbital moment preferentially excites spin-up electrons, while

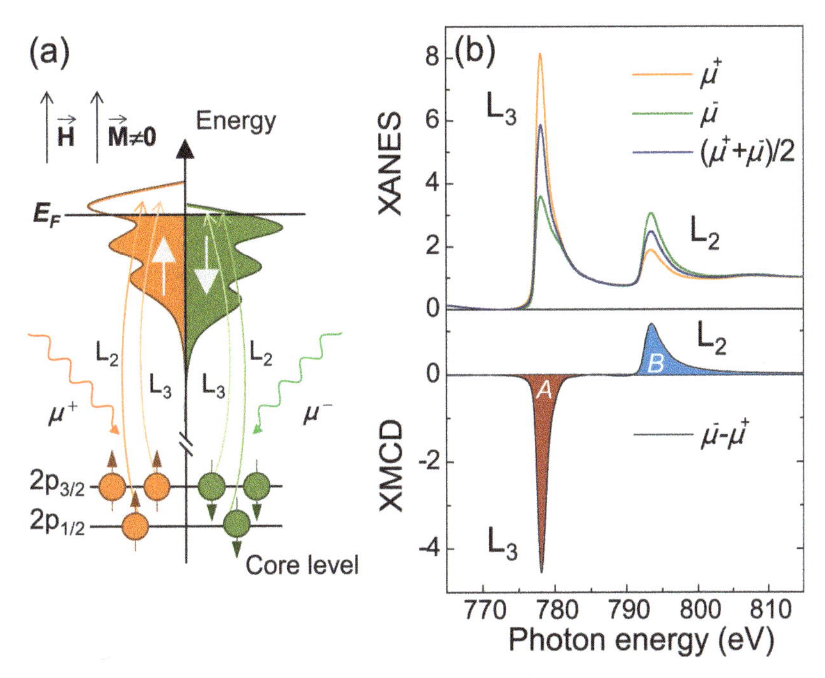

Figure 3.46. (a) Schematic diagram showing the electronic transitions involved in XMCD. (b) Results of XMCD in bulk Co. Reprinted from van der Laan and Figueroa (2014), copyright (2014), with permission from Elsevier.

radiation with helicity antiparallel to this moment preferentially excites spin-down electrons.

Figure 3.46(a) shows a schematic energy-level diagram of a ferromagnetic 3d metal together with photoionisation transitions at the L_2 and L_3 edges, corresponding to the thresholds for ionising the $2p_{1/2}$ and $2p_{3/2}$ core levels. In these materials the exchange interaction leads to an offset in energy of the d-band valence states depending on their electron spin orientation relative to the direction of an imposed magnetic field. As a result, there is a majority of occupied states of one spin orientation and a majority of unoccupied states of the opposite spin orientation. The minority spin direction is the direction of the magnetic field in figure 3.46(a). The 2p core states are spin–orbit split into two states depending on the relative orientation of the spin and angular momentum moments; these are parallel in the $2p_{3/2}$ state ($j = l + s = 1 + ½$) and antiparallel in the $2p_{1/2}$ state ($j = l - s = 1 - ½$). The dipole selection rule ($\Delta l = \pm 1$) for photoabsorption means that a transition from a *p*-state must be to *s* and *d* final states, but the cross-section for transitions to the $l + 1$ (*d*–) state generally dominate, so L-edge XANES is effectively a probe of the unoccupied part of the 3d valence band.

Figure 3.46(b) presents the results of experimental XANES measurements from cobalt, showing sharp peaks at the L_2 and L_3 edges. Historically, these peaks were commonly referred to as 'white lines' from the time when such spectra were recorded on photographic emulsions. The integrated intensity of these absorption peaks is proportional to the number of unoccupied 3d states, N:

$$I_{L_2} + I_{L_3} = CN,$$

where C is the modulus squared of the matrix element for $p \rightarrow d$ transitions. In general, this equation is only valid if the XANES intensities are averaged over all incidence angles, but for bulk fcc and bcc solids the high crystalline symmetry removes this requirement. Using incident circularly polarised radiation leads to selectivity in the spin orientation of the excited electrons. Specifically, if the helicity vector of the radiation is parallel to the 2p orbital moment spin-up electrons are preferentially emitted; if these directions are antiparallel, spin-down electrons are preferentially emitted. The relative intensities of the L_3 and L_2 adsorption edge peaks are thus determined by the number of unoccupied 3d states (holes) of the appropriate spin. The opposite spin orientations of the $2p_{3/2}$ (L_3 edge) and the $2p_{1/2}$ (L_2 edge) allow circularly polarised radiation to probe empty states of the opposite spin; positive helicity of the incident radiation, μ^+, will enhance the L_3 edge peak, exciting 62.5% spin-up electrons, while radiation of negative helicity, μ^-, will enhance the L_2 edge peak, exciting 75% spin-down electrons. Circular dichroism is defined as the difference between the absorption using the two opposite light helicities, so the XMCD signal will be negative at one absorption edge and positive at the other. Important sum rules allow the orbital and spin magnetic moments of the absorbing atoms to be extracted separately from the XMCD measurements. For isotropic materials, measured in a magnetically saturated sample with a strong external field along the direction of the X-ray propagation, the spin moment is

$$m_s = -\frac{(A - 2B)}{C}\mu_B,$$

while the orbital moment along the field direction is

$$m_o = -\frac{2(A + B)}{3C}\mu_B,$$

in units of Bohr magnetons, μ_B. A and B are the XMCD signals at the two L edges (figure 3.46(b)). For anisotropic materials, including ultra-thin films and surfaces, more complex orientation-dependent sum-rule equations are involved as given in a similar format, for example, in the review by Stöhr (1999).

The ability of XMCD to provide atomic element-specific magnetic moments, separated into orbital and spin contributions, makes the technique extremely valuable in studies of bulk materials but also surfaces and thin films, including multilayer structures. An interesting surface science application was in the study of the Cu(100)c(2 × 2)-Mn surface-alloy phase formed by deposition of manganese onto a Cu(100) surface, the Mn atoms occupying substitutional sites as shown in figure 3.47. This surface phase has no equivalent bulk Cu-Mn alloy phase. A quantitative structural study of the surface alloy by Wuttig *et al* (1993) revealed that the Mn atoms lie some 0.3 Å higher above the surface than the surrounding Cu

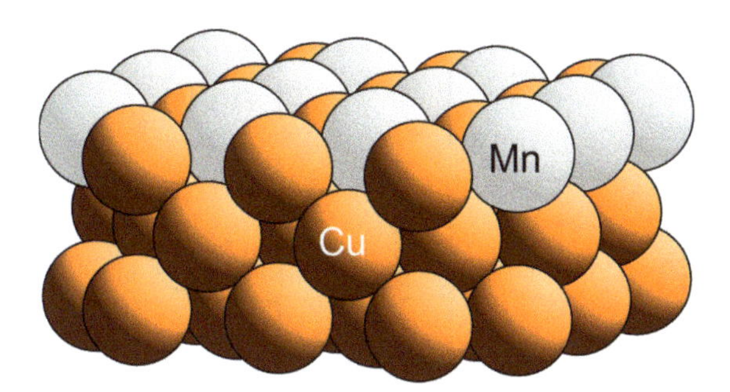

Figure 3.47. Schematic ball model of the Cu(100)c(2 × 2)-Mn surface-alloy structure.

atoms in the alloy layer, implying that the effective radii of the Mn atoms are much larger than those of the Cu atoms, or of Mn atoms in bulk Mn. Based on the results of *ab initio* total energy calculations, these authors concluded that this effect is a consequence of the Mn atoms being magnetic with a magnetic moment of 3.64 μ_B. The calculations favoured the Mn atoms being ferromagnetic, although the magnetic ordering was not found to be relevant to the buckling effect.

Mn $L_{2,3}$ XANES measurements by O'Brien and Tonner (1995) were found to be consistent with this predicted high-spin state from the Mn atoms in this system, but XMCD measurements to probe the magnetic ordering found no effect at room temperature, clearly inconsistent with ferromagnetic ordering. Later XMCD measurements confirmed the lack of XMCD signal at room temperature, but did find a small spin moment at 43 K of 0.12 ± 0.01 μ_B, falling to 0.03 ± 0.01 μ_B at 126 K, with no detectable orbital component of the moment (Kimura *et al* 2007). Above ~40 K these authors inferred that the surface alloy is in a paramagnetic phase.

XMCD has been widely used in the study of thin films and artificial multilayer structures, such as those in investigations and applications of giant magnetoresistance (or GMR; Parkin 1995). An interesting case is in studies of Co–Cu multilayers in which Cu $L_{2,3}$-edge XMCD shows that the Cu atoms acquire a local *d*-spin magnetic moment parallel to the Co moment, albeit with values of $\lesssim$0.1 μ_B, more than an order of magnitude less than that on the Co atoms (Held *et al* 1996). The effect is largest for the thinnest Cu layers in which most Cu atoms are close to the interfaces. XMCD has also revealed the effect of chemisorption on thin ferromagnetic films. For example, CO adsorption at 200 K onto six monolayers of Co grown epitaxially on Pd(111) was found to cause a spin reorientation of the magnetisation from parallel to the surface to perpendicular to the surface (Matsumura *et al* 2002). A further surface science application of XMCD is in investigations of the magnetic properties of nanoscale clusters. One interesting example of such a sample is a cluster of Au with its surface passivated by adsorption of an organic molecule such as a thiol; low-temperature XMCD studies have shown there to be a magnetic moment on the Au atoms (e.g., Negishi *et al* 2006 and references therein). Figure 3.48 shows the results of an investigation by Negeshi *et al* (2006) of Au clusters terminated by the thiol glutathione, specifically $Au_{18}(GS)_{14}$, at 2.7 K in a

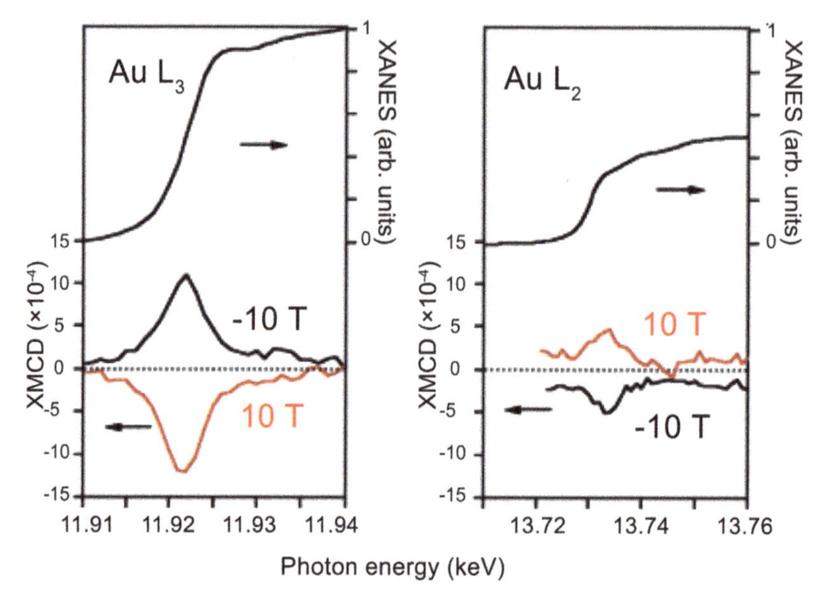

Figure 3.48. Au L_3 and L_2 XANES and MXCD with an applied field of 10 T from $Au_{18}(SG)_{14}$ at 2.7 K. The scale of the XMCD signal is relative to the XANES edge jump. Reprinted with permission from Negishi *et al* (2006). Copyright (2002) American Chemical Society.

magnetic field of 10 T. The temperature dependence of the XMCD signal is consistent with paramagnetism. The origin of this effect is debated, but is presumably related in some way to the modification of the electron structure of the Au atoms bonded to the thiol, although it also seems to be specific to Au in cluster form. The effect has been confirmed by SQUID measurements and has also been observed in nanoparticles of platinum, palladium and zinc oxide, reviewed by Nealon *et al* (2012) and by Crespo *et al* (2013).

3.7.3 X-ray linear dichroism

Magnetic materials also show linear dichroism, i.e., a difference in absorption of linearly polarised radiation depending on whether the polarisation vector is parallel or perpendicular to the local magnetic moment. Ferromagnetism can therefore be investigated by both XMCD and XMLD, but the signal in XMLD is very significantly weaker than in XMCD (e.g., Schwickert *et al* 1998). However, a key difference in these techniques is that while XMCD gives no signal in an antiferromagnet, being proportional to the magnetic moment $\langle M \rangle$, XMLD is proportional to $\langle M^2 \rangle$ and so antiferromagnetism can be investigated with XMLD (van der Laan *et al* 1986). X-ray linear dichroism (XLD) is also sensitive to ferroelectric behaviour (e.g., Stöhr *et al* 1999, Polisetty *et al* 2012). Evidently, XLD measurements require the control over the direction of linear polarisation of incident X-rays that is provided by synchrotron radiation. Some examples of spatially resolved XLD measurements are described in chapter 5.

References

Avila J, Casado C and Asensio M C *et al* 1995 Bulk Fermi surface determination by tuning the photoelectron kinetic energy *J. Vac. Sci. Technol.* A **13** 1501

Bauer E 2012 LEEM and UHV-PEEM: a retrospective *Ultramicroscopy* **119** 18–23

Baumberger F, Greber T and Osterwalder J 2001 Fermi surfaces of the two-dimensional surface states on vicinal Cu(111) *Phys. Rev.* B **64** 195411

Bentmann H, Maaß H and Braun J *et al* 2021 Profiling spin and orbital texture of a topological insulator in full momentum space *Phys. Rev.* B **103** L161107

Berens J, Bichelmaier S and Fernando N K *et al* 2020 Effects of nitridation on SiC/SiO_2 structures studied by hard X-ray photoelectron spectroscopy *J. Phys.: Energy* **2** 035001

Beutler A, Lundgren E and Nyholm R *et al* 1998 Coverage- and temperature-dependent site occupancy of carbon monoxide on Rh(111) studied by high-resolution core-level photoemission *Surf. Sci.* **396** 117–36

Boswick A, McChesney J and Ohta T *et al* 2009 Experimental studies of the electronic structure of graphene *Prog. Surf. Sci.* **84** 380

Borg M, Birgersson M, Smedh M, Mikkelsen A, Adams D L, Nyholm R, Almbladh C O and Andersen J N 2004 Experimental and theoretical surface core-level shifts of aluminium (100) and (111) *Phys. Rev.* B **69** 235418

Bradshaw A M and Woodruff D P 2015 Molecular orbital tomography for adsorbed molecules: is a correct description of the final state really unimportant? *New J. Phys.* **17** 013033

Broekman L, Tadich A and Huwald E *et al* 2005 First results from a second generation toroidal electron spectrometer *J. Elect. Spect. Rel. Phenom.* **144–147** 1001–4

Brookes N B, Clarke A, Johnson P D and Weinert M 1990 Magnetic surface states on Fe(001) *Phys. Rev.* B **41** 2643

Bunău O and Joly Y 2009 Self-consistent aspects of X-ray absorption calculations *J. Phys.: Condens. Matter* **21** 345501

Chuang C W *et al* 2021 Resonant photoemission spectroscopy of the ferromagnetic Kondo system $CeAgSb_2$ *Electron Struct* **3** 034001

Crespo P, de la Presa P and Marín P *et al* 2013 Magnetism in nanoparticles: tuning properties with coatings *J. Phys.: Condens. Matter* **25** 484006

Kalha C *et al* 2021 Hard X-ray photoelectron spectroscopy: a snapshot of the state-of-the-art in 2020 *J. Phys. Condens. Matter* **33** 233001

Kim J W, Dil J H, Kampen T and Horn K 2007 Circular dichroism in photoelectrons from adsorbed chiral molecules *AIP Conf. Proc.* **879** 1607

Dedkov Y S, Fonin M, Rüdiger U and Güntherodt G 2006 Spin-resolved photoelectron spectroscopy of the MgO/Fe(110) system *Appl. Phys.* A **82** 489

Dubs R L, Dixit S N and McKoy V 1985 Circular dichroism in photoelectron angular distributions from adsorbed atoms *Phys. Rev.* B **32** 8389–91

Duncan D A, Kreikemeyer-Lorenzo D and Primorac E *et al* 2014 V-doped TiO2(110): quantitative structure determination using energy-scanned photoelectron diffraction *Surf. Sci.* **630** 64–70

Feder R, Jennings P J and Jones R O 1976 Spin-polarization in LEED: a comparison of theoretical predictions *Surf. Sci.* **61** 307–16

Fredriksson W, Malmgren S, Gustafsson T, Gorgoi M and Edström K 2012 Full depth profile of passive films on 316L stainless steel based on high resolution HAXPES in combination with ARXPS *Appl. Surf. Sci.* **258** 5790–7

Gadzuk J W 1974 Surface molecules and chemisorption. II. Photoemission angular distributions *Phys. Rev.* B **10** 5030–44

George G N and Gorbaty M L 1989 Sulfur K-edge X-ray absorption spectroscopy of petroleum asphaltenes and model compounds *J. Am. Chem. Soc.* **111** 3182–6

Gibson J S, Narayanan S and Swallow J E N *et al* 2022 Gently does it!: *in situ* preparation of alkali metal–solid electrolyte interfaces for photoelectron spectroscopy *Faraday Discuss.* **236** 267

Giordanino F, Borfecchia E and Lomachenko K A *et al* 2014 Interaction of NH_3 with Cu-SSZ-13 catalyst: a complementary FTIR, XANES, and XES study *J. Phys. Chem. Lett.* **5** 1552–9

Goldberg S M, Fadley C S and Kono S 1978 Photoelectric cross-sections for fixed-orientation atomic orbitals: relationship to the plane-wave final state approximation and angle-resolved photoemission *Solid State Commun.* **28** 459–63

Gorgoi M *et al* 2009 The high kinetic energy photoelectron spectroscopy facility at BESSY progress and first results *Nucl. Instrum. Methods* A **601** 48–53

Grobman W D 1978 Angle-resolved photoemission from molecules in the independent-atomic-center approximation *Phys. Rev.* B **17** 4573–85

Grönbeck H, Klacar S and Martin N M *et al* 2012 Mechanism for reversed photoemission core-level shifts of oxidized Ag *Phys. Rev.* B **85** 115445

Haags A *et al* 2022 Momentum space imaging of sigma orbitals for chemical analysis *Sci. Adv.* **8** eabn0819

Held G A, Samant M G and Stöhr J *et al* 1996 X-ray magnetic circular dichroism study of the induced spin polarization of Cu in Co/Cu and Fe/Cu multilayers *Z. Phys.* B **100** 335–41

Hoesch M, Muntwiler M and Petrov V N *et al* 2004 Spin structure of the Shockley surface state on Au(111) *Phys. Rev.* B **69** 241401

Jablonski A and Tougaard C J 1998 Practical correction formula for elastic electron scattering effects in attenuation of Auger electrons and photoelectrons *Surf. Interface Anal.* **26** 17–29

Jablonski A and Powell C J 2009 Practical expressions for the mean escape depth, the information depth, and the effective attenuation length in Auger-electron spectroscopy and x-ray photo-electron spectroscopy *J. Vac. Sci. Technol. A* **27** 253–61

Jennings P J 1974 Spin-polarisation in LEED *Jpn. J. Appl. Phys Suppl.* **2** *Pt. 2* 661–6

Johnson P D 1997 Spin-polarized photoemission *Rep. Prog. Phys.* **60** 1217–304

Johnson P D and Güntherodt G 2007 Handbook of magnetism and advanced magnetic materials ed H Kronmuller and S Parkinvol. 3 *Novel Techniques for Characterizing and Preparing Samples* (New York: Wiley)

Joyner R W, Roberts M W and Yates K 1979 A 'high-pressure' electron spectrometer for surface studies *Surf. Sci.* **87** 501–9

Kera S, Tanaka S and Yamane H *et al* 2006 Quantitative analysis of photoelectron angular distribution of single-domain organic monolayer film: NTCDA on GeS(0 0 1) *Chem. Phys.* **325** 113–20

Kevan S D and Gaylord R H 1987 High-resolution photoemission study of the electronic structure of the noble-metal (111) surfaces *Phys. Rev.* B **36** 5809–18

Kimura A, Asanao S and Kambe T *et al* 2007 Electron correlation and magnetic properties of c (2 × 2)CuMn/Cu(001) two-dimensional surface alloys *Phys. Rev.* B **76** 115416

Kirschner J and Feder R 1979 Spin polarization in double diffraction of low-energy electrons from W(001): experiment and theory *Phys. Rev. Lett.* **42** 1008–11

Knapp J A, Himpsel F J and Eastman D E 1979 Experimental energy band dispersions and lifetimes for valence and conduction bands of copper using angle-resolved photoemission *Phys. Rev.* B **19** 4952–64

Larsson F, Keller J and Olsson J *et al* 2020 Amorphous tin-gallium oxide buffer layers in (Ag,Cu)(In,Ga)Se_2 solar cells *Solar Energy Mat. Sol. Cells* **215** 110647

LaShell S, McDougall B A and Jensen E 1996 Spin splitting of an Au(111) surface state band observed with angle resolved photoelectron spectroscopy *Phys. Rev. Lett.* **77** 3419–2422

Lin C, Ochi M and Noguchi R *et al* 2021 Visualization of the strain-induced topological phase transition in a quasi-one-dimensional superconductor $TaSe_3$ *Nat. Mat.* **20** 1093–9

Lüftner D, Ules T and Reinisch E M *et al* 2013 Imaging the wave functions of adsorbed molecules *Proc. Nat. Acad, Sci.* **111** 605–10

Matsui F, Makita S and Matsuda H *et al* 2020 Photoelectron momentum microscope at BL6U of UVSOR-III synchrotron *Jpn. J. Appl. Phys.* **59** 067001

Matsumura D, Yokoyama T and Amemiya K *et al* 2002 X-ray magnetic circular dichroism study of spin reorientation transitions of magnetic thin films induced by surface chemisorption *Phys. Rev.* B **66** 024402

Mudd J J, Lee T L and Muñoz-Sanjosé V *et al* 2014 Valence-band orbital character of CdO: a synchrotron-radiation photoelectron spectroscopy and density functional theory study *Phys. Rev.* B **89** 165305

Nealon G L, Donnio B and Greget R *et al* 2012 Magnetism in gold nanoparticles *Nanoscale* **4** 5244–58

Negishi Y, Tsunoyama H and Suzuki M *et al* 2006 X-ray magnetic circular dichroism of size-selected, thiolated gold clusters *J. Am. Chem. Soc.* **128** 12034

Offenbacher H, Lüftner D and Ules T *et al* 2015 Orbital tomography: molecular band maps, momentum maps and the imaging of real space orbitals of adsorbed molecules *J. Electr. Spectrosc. Rel. Phenon.* **204** 92–101

Ohta T, Bostwick A, Seyller T, Horn K and Rotenberg E 2006 Controlling the electronic structure of bilayer graphene *Science* **313** 951–4

Okuda T, Takeichi Y and Maeda Y *et al* 2008 *Rev. Sci. Instrum.* **79** 123117

Okuda T, Miyamaoto K and Miyahara H *et al* 2011 Efficient spin resolved spectroscopy observation machine at Hiroshima synchrotron radiation center *Rev. Sci. Instrum.* **82** 103302

O'Brien W L and Tonner B P 1995 Magnetic properties of Mn/Cu(001) and Mn/Ni(001)c(2 X2) surface alloys *Phys. Rev. B.* **51** 617–8

O'Neill M R, Kalisvaart M, Dunning F B and Walters G K 1975 Electron-spin polarization in low-energy electron diffraction from tungsten (001) *Phys. Rev. Lett.* **34** 1167–117

Osterwalder J, Greber T, Wetli E, Wider J and Neff H J 2000 Full hemispherical photoelectron diffraction and Fermi surface mapping *Prog. Surf. Sci.* **64** 65–87

Parkin S S P 1995 Giant magnetoresistance in magnetic nanostructures *Annu. Rev. Mater. Sci.* **25** 357–88

Pippard A B 1957 An experimental determination of the fermi surface in copper *Phil Trans. Roy. Soc.* A **250** 325–57

Polcik M, Allegretti F and Sayago D I *et al* 2004 Circular dichroism in core level photoemission from an adsorbed chiral molecule *Phys. Rev. Lett.* **92** 236103

Polisetty S, Zhou J and Karthik J *et al* 2012 X-ray linear dichroism dependence on ferroelectric polarization *J. Phys. Condens. Matter* **24** 245902

Powell C J and Jablonski A 2011 *NIST Electron Effective-Attenuation-Length Database, Version 1.3, SRD 82* (Gaithersburg, MD: National Institute of Standards and Technology)

Puschnig P, Berkebile S and Fleming A J *et al* 2009 Reconstruction of molecular orbital densities from photoemission data *Science* **326** 702–6

Razado-Colambo I, Avila J and Vignaud D *et al* 2018 Structural determination of bilayer graphene on SiC(0001) using synchrotron radiation photoelectron diffraction *Sci. Rep.* **8** 10190

Razzoli E, Sassa Y and Drachuck G *et al* 2010 The Fermi surface and band folding in $La_{2-x}Sr_xCuO_4$, probed by angle-resolved photoemission *New J. Phys.* **12** 125003

Rehr J J, Kas J J, Vila F D, Prange M P and Jorissen K 2010 Parameter-free calculations of X-ray spectra with FEFF9 *Phys. Chem. Chem. Phys.* **12** 5503–13

Rotenberg E, Denlinger J D and Kevan S D *et al* 1996 Fermi surface mapping using a third generation light source *MRS Online Proc. Lib.* **437** 47–52

Rubio-Zuazo J and Castro G R 2011 Effective attenuation length dependence on photoelectron kinetic energy for Au from 1 keV to 15 keV *J. Electron Spectrosc. Rel. Phenom.* **184** 384–90

Salmeron M and Schlögl R 2008 Ambient pressure photoelectron spectroscopy: a new tool for surface science and nanotechnology *Surf. Sci. Rep.* **63** 169–99

Sano M, Komorita S and Yamatera H 1992 XANES spectra of copper(II) complexes: correlation of the intensity of the 1s → 3d transition and the shape of the complex *Inorg. Chem.* **31** 459–63

Scheinfein M R, Ungaris J, Kelley M H, Pierce D T and Celotta R J 1990 Scanning electron microscopy with polarization analysis (SEMPA) *Rev. Sci. Instrum.* **61** 2501–27

Schröder K, Prinz G A, Walker K-H and Kisker E 1985 Spin- and angle-resolved photoemission study of (110) Fe films grown on GaAs by molecular beam epitaxy *J. Appl. Phys.* **57** 3669–71

Schwickert M M, Guo G Y, Tomaz M A, O'Brien W L and Harp G R 1998 X-ray magnetic linear dichroism in absorption at the L edge of metallic Co, Fe, Cr, and V *Phys. Rev.* B **58** R4289–4292

Seah M P and Gilmore I S 2001 Simplified equations for correction parameters for elastic scattering effects in AES and XPS for Q, β and attenuation lengths *Surf. Interface Anal.* **31** 835–46

Shirley D A 1973 ESCA *Adv. Chem. Phys.* **13** 85

Smedh M, Beutler A and Ramsvik T *et al* 2001 Vibrationally resolved C 1s photoemission from CO absorbed on Rh(1 1 1): the investigation of a new chemically shifted C 1s component *Surf. Sci.* **491** 99–114

Smith R J, Anderson J and Lapeyre G J 1976 Adsorbate orientation using angle-resolved polarization-dependent photoemission *Phys. Rev. Lett.* **37** 1081–4

Smith N V, Benbow R L and Hurych Z 1980 Photoemission spectra and band structures of d-band metals. VIII. Normal emission from Cu(111) *Phys. Rev. B.* **21** 4331–6

Sobota J A, He Y and Shen Z-X 2021 Angle-resolved photoemission studies of quantum materials *Rev. Mod. Phys.* **93** 025006

Stöhr J 1999 Exploring the microscopic origin of magnetic anisotropies with X-ray magnetic circular dichroism (XMCD) spectroscopy *J. Magn. Magn. Mater.* **200** 470–97

Stöhr J, Scholl A and Regan T J *et al* 1999 Images of the antiferromagnetic structure of a NiO (100) surface by means of X-ray magnetic linear dichroism spectromicroscopy *Phys. Rev. Lett.* **83** 1862–5

Swallow J E N, Fraser M W and Kneusels N-J H *et al* 2022 Revealing solid electrolyte interphase formation through interface-sensitive operando X-ray absorption spectroscopy *Nat. Commun.* **13** 6070
Tamai A, Meevasana W and King P D C *et al* 2013 Spin–orbit splitting of the Shockley surface state on Cu(111) *Phys. Rev.* B **87** 075113
Tillborg H, Nilsson A and Mårtensson N 1992 Studies of the CO–H,H,-Ni(100) system using photoelectron spectroscopy *Surf. Sci.* **273** 47–60
Tillmann D, Thiel R and Kisker E 1989 Very-low-energy spin-polarized electron diffraction from Fe(001) *Z. Phys.* B **77** 1–2
Timoshenko J, Halder A and Yang B *et al* 2018 Subnanometer substructures in nanoassemblies formed from clusters under a reactive atmosphere revealed using machine learning *J. Phys Chem.* C **122** 21686
Trotochaud L, Head A R, Karslioğlu O, Kyhl L and Bluhm H 2017 Ambient pressure photoelectron spectroscopy: practical considerations and experimental frontiers *J. Phys. Condens. Matter* **29** 053002
Tusche C, Krasyuk A and Kirschner J 2015 Spin resolved bandstructure imaging with a high resolution momentum microscope *Ultramicroscopy* **159** 520–9
Ueda S and Sakuraba Y 2021 Direct observation of spin-resolved valence band electronic states from a buried magnetic layer with hard X-ray photoemission *Sci. Technol. Adv. Mats.* **22** 317
Ueno N, Kitamura A and Okudaira K K *et al* 1997 Angle-resolved ultraviolet photoelectron spectroscopy of thin films of bis(1,2,5-thiadiazolo)-*p*-quinobis (1,3-dithiole) on the MoS_2 surface *J. Chem. Phys.* **107** 2079–88
Valla T, Fedorov A V and Johnson P D *et al* 1999 Evidence for quantum critical behavior in the optimally doped cuprate $Bi_2Sr_2CaCu_2O_{8+\delta}$ *Science* **285** 2110–3
van der Laan G, Thole B T and Sawatzky G A *et al* 1986 Experimental proof of magnetic X-ray dichroism *Phys. Rev.* B **34** 6529–31
van der Laan G and Figueroa A L 2014 X-ray magnetic circular dichroism—a versatile tool to study magnetism *Coord. Chem. Rev.* **277–278** 95–129
Williams A R and Lang N D 1977 Atomic chemisorption on simple metals: Chemical trends and core-hole relaxation effects *Surf. Sci.* **68** 138–48
Wuttig M, Gauthier Y and Blügel S 1993 Magnetically driven buckling and stability of ordered surface alloys: Cu(100)$c(2 \times 2)$Mn *Phys. Rev. Lett.* **70** 3619–22
Yamamoto S, Bluhm H and Andersson K *et al* 2008 *In situ* X-ray photoelectron spectroscopy studies of water on metals and oxides at ambient conditions *J. Phys.: Condens. Matter* **20** 184025

IOP Publishing

Surface Science and Synchrotron Radiation

Phil Woodruff

Chapter 4

Geometrical structure: diffraction of X-rays and photoelectrons

Methods that rely on the use of synchrotron radiation to obtain quantitative information on the structure of surfaces are described, with a number of examples of their application. Surface X-ray diffraction (SXRD) is a variant of standard X-ray diffraction (XRD) for the determination of bulk structures, but the use of specific scattering geometries allows the weak scattering from surface layers to be separated from the much stronger scattering signal from the underlying bulk. Surface structural information can also be extracted by measuring photoabsorption in X-ray standing waves created by XRD in the underlying bulk. Alternative methods rely not on X-ray scattering of surface atoms but elastic scattering of photoelectrons created by the incident photons. Specifically, these techniques include both near-edge and extended X-ray absorption fine structure, but also photoelectron diffraction. Using photoelectron detection these techniques provide local element-specific and chemical-state-specific quantitative structural information.

4.1 Introduction

X-ray diffraction (XRD) is the core technique for determining the structure of three-dimensionally periodic (crystal) structures and is applied successfully to a wide range of complex materials including macromolecular crystals. It is one of the most widely exploited of synchrotron radiation techniques. However, the fact that X-rays are scattered weakly by atoms and penetrate deep into solids means that XRD is not intrinsically surface specific. Instead, the benchmark diffraction technique for determining surface structures is low-energy electron diffraction (LEED), a technique for which the source is a simple low-energy electron gun readily available in most surface scientists' home laboratories. The underlying physical principles of XRD and LEED are the same, namely the coherent interference of elastic scattering

doi:10.1088/978-0-7503-3847-9ch4

by the atoms in the material. The translational periodicity of the structure ensures that the scattered X-rays or electrons are emitted in well-defined directions as 'diffracted beams', and analysis of these directions allows one to determine the periodicity, or unit cell (or unit mesh in the case of a surface), of the sample. The intensities of these diffracted beams are determined by the relative positions of atoms within these unit cells, so a full structure determination is achieved by analysing these intensities. A key difference between XRD and LEED, however, is in the magnitude of the cross-sections for elastic and inelastic scattering. X-ray scattering is weak, so it penetrates deep into a target solid, making it ideal to determine the structure of three-dimensionally periodic structures. By contrast, both elastic and inelastic scattering cross-sections for low-energy electrons (typically <500 eV) are as much as 10^6 times larger, leading to elastically scattered electrons in this energy range only sampling the outermost few atomic layers of a solid (see chapter 1). LEED is therefore intrinsically surface specific and is a natural complement for surface studies to XRD for bulk crystalline structure determination. However, the large cross-section for elastic scattering does introduce a complication, namely that the diffracted intensities are strongly influenced by multiple scattering, making their analysis much more computationally demanding. As a result, there can be advantages to performing XRD from surfaces if the experiment can be arranged in such a way that the weak surface scattering can be detected without the signal being swamped by the much stronger scattering of the X-rays from the underlying bulk crystal. Ways of achieving this to establish a viable surface X-ray diffraction technique (SXRD) are described in section 4.2.

A quite different pair of surface structural techniques that exploit the special properties of synchrotron radiation, notably the ability to scan the photon energy in the X-ray energy range, are actually based not on X-ray scattering but on low-energy electron scattering. Instead of using an external electron gun, as in LEED, these methods are based on elastic scattering of low-energy photoelectrons generated by the incident X-rays. These two techniques, namely photoelectron diffraction (PhD) and SEXAFS, a surface-sensitive version of extended X-ray absorption fine structure (EXAFS), both involve the detection of the effects of the coherent interference of elastically scattered electrons, but whereas LEED involves scattering of an incident plane wave, PhD and EXAFS use an incident spherical wave centred on the photo-emitting atoms. This leads to techniques that detect the *local* structure around the photoemitting atoms rather than the long-range ordered structure detected in LEED (and XRD). These techniques are described in detail in section 4.3.

There is one further synchrotron radiation surface structural technique that is based on XRD in the bulk of a crystal. The X-ray standing wave technique (XSW) exploits the fact that at an X-ray Bragg diffraction condition the interference of the incident and scattered waves creates a standing wave with a periodicity equal to the spacing of Bragg diffraction planes. Fine adjustments of the exact diffraction conditions around the nominal Bragg condition cause this standing wave to shift in a controlled fashion relative to the atomic scatterers in the crystal, so monitoring the absorption of the standing wave at atoms in the crystal as its phase is scanned allows one to determine the relative positions of the absorbing atoms. Surface specificity in the XSW technique is achieved by using photoemission to detect the local photoabsorption, the inelastic scattering of these low-energy photoelectrons ensuring that this measured signal originates only from the outermost few atomic layers, as in standard photoemission experiments. The application of XSW to determine surface structural information is described more fully in section 4.4.

4.2 Surface X-ray diffraction

A key feature of any surface diffraction technique, be it of incident X-rays or electrons (or atoms), is to recognise that while XRD of bulk crystals exploits the three-dimensional periodicity of crystals, the surface of a crystalline surface is only two-dimensionally periodic (parallel to the surface). The existence of a surface intrinsically terminates the periodicity perpendicular to the surface, but surface specificity in the measurement technique also ensures that the signal from successive atomic layers close to the surface will not be equivalent. Of course, it is also true that even for a clean surface of an elemental crystal there is some relaxation of the interlayer spacings close to the surface. To understand the implications of the reduced periodicity for an XRD experiment, it is useful to go back to the basics of X-ray crystallography. As in all elastic scattering experiments, the key equations describing the process can be derived from the conditions for conservation of energy and momentum. An important feature of a three-dimensionally periodic crystal is that the momentum recoil of the solid must always correspond to the value of a reciprocal lattice vector, $\mathbf{g}_{hkl}$.

If the incident electron or photon beam, with a wavevector $\mathbf{k}_i$, leads to a scattered wavevector $\mathbf{k}_s$, then conservation of energy leads to

$$k_s^2 = k_i^2.$$

While conservation of momentum gives

$$\mathbf{k}_s = \mathbf{k}_i + \mathbf{g}_{hkl}.$$

The reciprocal lattice vector is defined as

$$\mathbf{g}_{hkl} = h\mathbf{a}^* + k\mathbf{b}^* + l\mathbf{c}^*.$$

The primitive translation vectors of the reciprocal lattice are given in terms of the primitive translation vectors of the real-space lattice, **a**, **b** and **c**, by

$$\mathbf{a}^* = 2\pi\frac{\mathbf{b}\times\mathbf{c}}{V},\quad \mathbf{b}^* = 2\pi\frac{\mathbf{c}\times\mathbf{a}}{V},\quad \mathbf{c}^* = 2\pi\frac{\mathbf{a}\times\mathbf{b}}{V},\quad V = \mathbf{a}.\,\mathbf{b}\times\mathbf{c},$$

V being the volume of the unit cell.

A useful graphical representation of the consequences of these two conservation laws is the Ewald sphere construction in reciprocal space, shown in figure 4.1(a). To construct this diagram, one draws a vector $\mathbf{k}_i$ terminating at the origin of the reciprocal lattice, then draws a sphere of radius k centred on the start of the vector $\mathbf{k}_i$. Any intersections of this Ewald sphere with reciprocal lattice points then correspond to the conditions of diffracted beams given by a vector $\mathbf{k}_{hkl}$ from the

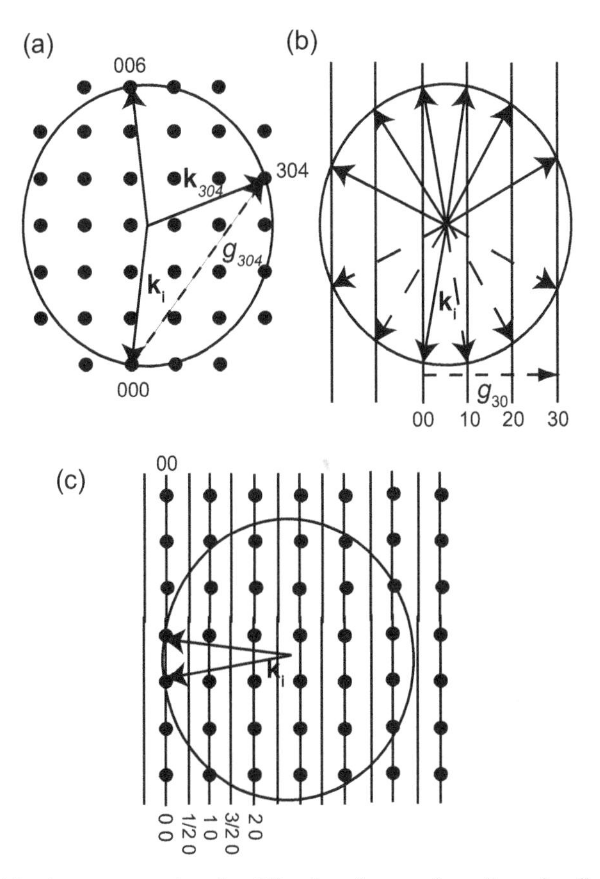

Figure 4.1. (a) Ewald sphere construction for diffraction from a three-dimensionally periodic structure. (b) Ewald sphere construction for diffraction from a two-dimensionally periodic surface. (c) As (b), but for the case of grazing incidence with the reciprocal lattice of the underlying three-dimensional periodic crystal superimposed on the crystal truncation rods.

centre of the sphere to the intersected reciprocal lattice point. Notice that as all these vectors have a length corresponding to the radius of the sphere, k, this ensures that energy conservation is obeyed, while the fact that both $\mathbf{k}_i$ and $\mathbf{k}_{hkl}$ end at reciprocal lattice points means that their difference is a reciprocal lattice vector $\mathbf{g}_{hkl}$, thereby ensuring momentum conservation.

In the equivalent experiment on a surface that is only two-dimensionally periodic, the momentum conservation constraint is relaxed perpendicular to the surface, with only the component of $\mathbf{k}$ parallel to the surface, $\mathbf{k}_{||}$, being conserved:

$$\mathbf{k}_{f||} = \mathbf{k}_{i||} + \mathbf{g}_{hk},$$

where $\mathbf{g}_{hk} = h\mathbf{a} + k\mathbf{b}$ is a reciprocal net vector with the primitive translation vectors of the reciprocal net being defined in terms of the primitive translation vector of the real-space net, $\mathbf{a}$ and $\mathbf{b}$, as $\mathbf{a}^* = 2\pi\frac{\mathbf{b}\times\mathbf{n}}{A}$, $\mathbf{b}^* = 2\pi\frac{\mathbf{n}\times\mathbf{a}}{A}$, $A = \mathbf{a}.\,\mathbf{b}\times\mathbf{n}$. A is the area of the unit mesh while $\mathbf{n}$ is a unit vector perpendicular to the surface.

The modified Ewald sphere construction for this situation is shown in figure 4.1(b). The loss of periodicity perpendicular to the surface in the real-space structure means that the discrete three-dimensionally periodic reciprocal lattice is replaced by a set of continuous 'rods' that pass through the reciprocal net points. Evidently, the condition for the Ewald sphere to intersect these rods is much less stringent than to intersect reciprocal lattice points, so the number of diffracted beams for any specific value of $\mathbf{k}_i$ is much larger. In figure 4.1(b) the forward-scattered (scattering angle <90°) and back-scattered (scattering angle >90°) beams are distinguished as dashed and continuous lines, respectively. This difference highlights a difference in the experimental data set required for structure determination of bulk crystals and their surfaces. In both cases, one needs to collect intensities of diffracted beams, but in the case of a bulk crystal these correspond to discrete values of h, k and l, although, as is clear from figure 4.1(a), it is necessary to change the direction (or modulus) of $\mathbf{k}_i$, in order to collect a large number of different (hkl) beams. This is typically achieved by rotating the crystalline target, although using a powdered sample leads to many different crystal orientations being sampled simultaneously.

In the case of diffraction from a surface the momentum transfer perpendicular to the surface, $l = \Delta\mathbf{k}_\perp$, is a continuous, rather than discrete, variable, so diffracted (hk) beam intensities are collected as 'rod scans' with $\mathbf{k}_{hk}$ terminating at different points on the reciprocal net rod. In the case of LEED this is easily achieved by simply varying the electron energy, which changes the radius of the Ewald sphere. In SXRD the detector geometry must be scanned, as shown schematically in figure 4.2. Details of the instrumentation at several synchrotron radiation facilities used to collect SXRD data are described in the book by Moritz and Van Hove (2022).

In most cases we expect a surface to be two-dimensionally periodic because the underlying bulk crystal is three-dimensionally periodic. One can define this two-dimensional periodicity of the bulk crystal, parallel to the surface, by a unit mesh

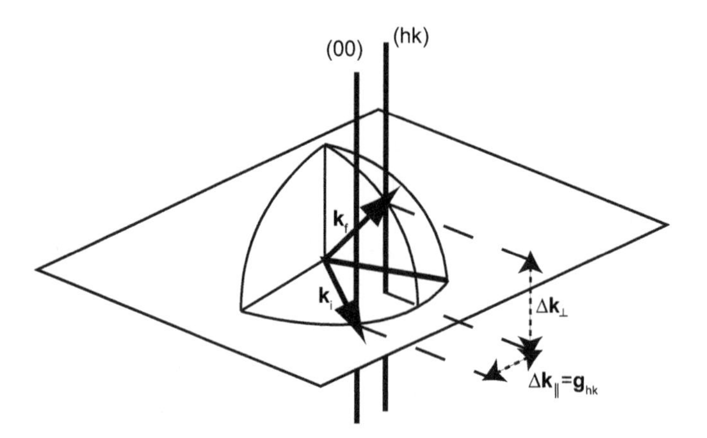

Figure 4.2. Schematic representation of the scattering geometry in a SXRD experiment showing how the Ewald sphere construction determining the direction of a (hk) diffracted beam as a function of $\Delta\mathbf{k}_\perp$ in a 'rod scan'.

defined by primitive translation vectors **a** and **b**. This unit mesh also describes the periodicity of a surface of an ideally bulk-terminated solid, but in most surfaces of interest the periodicity of the outermost atomic layers will be defined by a larger unit mesh with primitive translation vectors $\mathbf{a}'$ and $\mathbf{b}'$. This larger unit mesh may be a consequence of a reconstruction of the surface layer(s) of an otherwise ideally terminated crystal, but in most cases of interest this will be due to the presence of some kind of overlayer of adsorbed atoms or molecules. The surface and substrate primitive translation vectors can be related by

$$\mathbf{a}' = G_{11}\mathbf{a} + G_{12}\mathbf{b} \quad \text{and} \quad \mathbf{b}' = G_{21}\mathbf{a} + G_{22}\mathbf{b}.$$

If all of the four matrix coefficients G_{ij} are integers or rational fractions the surface is then *commensurate* with the underlying crystal. In this case, there exists a unit mesh with primitive translation vectors $n\mathbf{a}'$ and $m\mathbf{b}'$, with n and m both integers, that is common to both the surface and the underlying crystal. In most commensurate surfaces of interest the values of at least some of the G_{ij} coefficients are larger than unity, so the surface unit mesh is larger than the two-dimensional unit mesh of the underlying bulk crystal. Evidently, this means that the surface reciprocal unit mesh is smaller than that of the underlying bulk, leading to a smaller spacing of the reciprocal net rods and thus to 'extra' diffracted beams. These extra beams are labelled relative to the reciprocal mesh of the underlying bulk, and so have fractional indices. Figure 4.1(c) shows the Ewald sphere construction for such a situation, with 'extra' half-order surface rods. While figure 4.1(b) shows the situation when the incident radiation is near-normal to the surface (as in a LEED experiment), figure 4.1(c) shows the situation with grazing incidence to the surface (as also depicted in figure 4.2). Grazing incidence is generally used in SXRD to make the measurement relatively more surface specific. As shown in section 1.2, even a grazing incidence angle smaller than the condition for total external reflection does not ensure that the measurement is specific to only the outermost few atom layers, but it does very significantly reduce the amplitude of the scattered intensity that arises from the bulk, relative to that from the surface layers.

Superimposed in figure 4.1(c) on the reciprocal space rods associated with the surface unit mesh is also the reciprocal lattice of the underlying bulk crystal. This diagram allows us to gain some insight into the likely form of 'rod scan' diffracted beam intensity measurements, although it is important to recognise that the Ewald sphere construction only predicts the conditions for the existence of diffracted beams, but not their relative intensities. Nevertheless, it is clear from the figure that if one scans the diffracted beam condition along the (00) rod, as shown in figure 4.1(c), one must pass through some reciprocal lattice points when the bulk diffraction condition is met. Evidently, this must lead to a huge enhancement of the scattered intensity, as one detects coherent scattering from very many bulk layers as well as from the small number of surface layers. This effect must be a feature of scans along integral order rods, known as crystal truncation rods (CTRs). By contrast, scans along fractional-order rods (FORs) never pass through these bulk diffraction conditions. Notice that the fractional-order-diffracted beams have scattering contributions only from those atoms that have the periodicity of the larger surface unit mesh. As such, interpretation of these scattered intensities can provide information on the relative positions of atoms within the surface layers having the larger surface unit mesh, but provide no information on the registry of these layers relative to the underlying bulk. This extra information is contained only in the CTR scans. These statements rely on the scattering cross-sections being sufficiently small that one can ignore any contribution from multiple scattering; this situation is generally true for X-ray scattering (but not for low-energy electron scattering as in LEED).

A more quantitative description of the nature of a CTR scan for an ideally erminated crystalline solid can be derived from a simple single-scattering treatment of diffraction from a crystal of finite size N_1a × N_2b × N_3c. The scattering momentum transfer in terms of components in the directions defined by **a**, **b** and **c** can be written as $\Delta\mathbf{k} = \Delta\mathbf{k}_a + \Delta\mathbf{k}_b + \Delta\mathbf{k}_c$. The total scattering amplitude is then given by summation over unit cells:

$$A_{\Delta k} = \sum_{n_1=1}^{N_1}\sum_{n_2=1}^{N_2}\sum_{n_3=1}^{N_3} \exp[i(\Delta k_a n_1 a + \Delta k_b n_2 b + \Delta k_c n_3 c)]F_{\Delta k},$$

where $F_{\Delta k} = \sum_j f_j \exp(i\Delta\mathbf{k}\cdot\mathbf{r}_j)$ is the summation of atoms of scattering factor f_j and coordinates $\mathbf{r}_j$ within each unit cell.

The scattering intensity is then the modulus squared of this amplitude:

$$I_{\Delta k} = F_{\Delta k}{}^2 \frac{\sin^2\left(\frac{1}{2}N_1\Delta k_a a\right)}{\sin^2\left(\frac{1}{2}\Delta k_a a\right)} \frac{\sin^2\left(\frac{1}{2}N_2\Delta k_b b\right)}{\sin^2\left(\frac{1}{2}\Delta k_b b\right)} \frac{\sin^2\left(\frac{1}{2}N_3\Delta k_c c\right)}{\sin^2\left(\frac{1}{2}\Delta k_c c\right)}.$$

If the crystal is sufficiently large (nominally infinite) such that perfect three-dimensional periodicity can be assumed then, as stated above, $\Delta\mathbf{k} = \mathbf{g}_{\mathrm{hkl}}$, which leads to the three-dimensional Laue conditions, $\Delta k_a a = 2\pi h$, $\Delta k_b b = 2\pi k$ and $\Delta k_c c = 2\pi l$, whence

$$I_{hkl} = F_{hkl}^2 N_1^2 N_2^2 N_3^2.$$

Now consider the scattering from the surface of a large crystal. If the surface is truly two-dimensionally periodic (if N_1 and N_2 are both very large so that the lateral dimensions are large), the first two of these Laue conditions will be satisfied but, as discussed above, perpendicular to the surface the crystal is not truly periodic so the third Laue condition is not satisfied. Indeed, the presence of the surface means that the intensity of the incident X-rays must decrease as they penetrate the solid, even if one were to use normal incidence. This may occur because of absorption (inelastic scattering), but must also be a consequence of elastic scattering, because if X-rays are scattered out of the surface by successive layers of the crystal this also attenuates the incident beam. Suppose that the amplitude of the incident X-ray wavefield is attenuated with a characteristic decay length dc, i.e., the distance d is measured in units of the appropriate unit cell dimension perpendicular to the surface, c. The scattered amplitude can then be written as

$$A_{\Delta k} = \sum_{n_1}^{N_1}\sum_{n_2}^{N_2} \exp[\mathrm{i}(\Delta k_a n_1 a + \Delta k_b n_2 b)] \sum_{n_3}^{N_3} \exp(-n_3/d)\exp[\mathrm{i}(\Delta k_c n_3 c)],$$

and proceeding with the summation to N_1 and N_2 parallel to the surface, but to infinity (i.e., to convergence of the damped series) perpendicular to the surface yields an intensity expression when the first two Laue conditions are satisfied of

$$I_{hk} = F_{hk}{}^2 N_1{}^2 N_2{}^2 \frac{1}{\left\{[1 - \exp(-1/\mathrm{d})]^2 + 4\exp(-1/\mathrm{d})\sin^2\left(\frac{1}{2}\Delta k_c c\right)\right\}}.$$

The final term in this expression still peaks at the third Laue condition ($\sin(\frac{1}{2}\Delta k_c c) = 0$), but it also describes the finite scattered intensity along the CTRs away from this three-dimensional diffraction condition. This peak intensity has a value $I_{hk}(\max) = F_{hkl}^2 N_1^2 N_2^2 d^2$.

The value of N_3, the number of layers in the finite-sized three-dimensionally periodic crystal, is replaced by d, the effective number of contributing layers. An important result that may be derived from the expression for I_{hk} is the intensity midway between the conditions that satisfy the third Laue condition, $\sin(\frac{1}{2}q_3 c) = 1$. Here, the intensity from a truly three-dimensionally periodic crystal is zero, but in the presence of the surface this becomes $I_{hkl} = F_{hk}^2 N_1^2 N_2^2/4$. This corresponds to one quarter of the intensity (one half of the amplitude) of the scattering from a single atomic layer.

There are number of important conclusions regarding the SXRD technique that emerge from this analysis. First, as discussed in chapter 1, any technique designed to give detailed information on the surface, defined as the outermost few atomic layers of a solid, must be surface specific, i.e., the information obtained from the technique should be dominated by the contribution of this thin surface layer. In section 1.2 it was shown that even using a grazing incidence angle of around the critical value for total reflection, one may probe a depth of tens of nanometres, so in terms of interlayer spacings d may be of order 100. The intensity of the diffracted beams arising from bulk diffraction conditions in a CTR ('Bragg peaks') may therefore be $\sim 10^4$ times larger than the intensity midway between these three-dimensional diffraction conditions, but at these mid-points ('anti-Bragg' conditions) the intensity is roughly that to be expected for scattering from a single atomic layer. Any modification of the structure of the surface relative to that of the bulk, due to a shift or addition of a single atomic layer, will then introduce a scattering contribution of similar amplitude to that of the underlying solid, so the resulting coherent interference may produce a very large proportional change in the diffracted intensity. This is illustrated by the simple example of figure 4.3, which shows a calculated CTR from an ideal bulk-terminated Ag(111) surface and from this surface with the outermost layer spacing reduced, or increased, by 5%. Note that the ordinate of this graph is the structure factor, $F(hk)$, which is the square root of the intensity. This example shows clearly that a change in the outermost layer spacing of 0.1 Å produces huge changes in the scattered intensity of the (10) beam at values of the perpendicular momentum transfer, $\Delta k_\perp$,

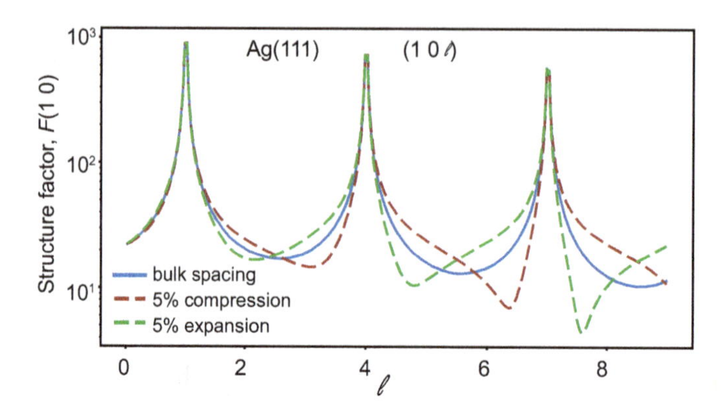

Figure 4.3. Calculated rod scans of the (10) beam diffracted from an ideally bulk-terminated Ag(111) surface (blue lines) and from the same surface with the outermost layer spacing reduced (red dashed line) and increased (green dashed line) by 5%. Courtesy of Phil Mousley, Diamond Light Source.

well-removed from the conditions for diffraction from the underlying bulk crystal, corresponding to integral values of l of 1, 4 and 7. CTR intensity measurements at these intermediate values of $\Delta k_{\perp}$ are clearly conditions for excellent surface specificity. The key to achieving surface specificity in SXRD is therefore to measure diffracted beam intensities in regions of Δk space in which bulk diffraction contributes little (or no) intensity. For CTRs this means measurements roughly midway between the conditions for diffraction from the underlying bulk. This condition is also met for all measurements of the fractional-order-diffracted beams.

While this clearly means that SXRD is a viable technique for determining the structure of surfaces under ultra-high vacuum (UHV) conditions, the benchmark technique for this purpose has been LEED, which has the advantage that it can be performed quite easily in any scientist's home laboratory without the need to gain access to synchrotron radiation. LEED does, however, have a significant disadvantage relative to SXRD in the data analysis and interpretation because of the much larger elastic scattering cross-sections for low-energy electrons relative to X-rays. While the ultimate method for extracting surface structures from both LEED and SXRD results relies on a trial-and-error approach in which the measured scattering intensities are compared with those of simulations for a range of possible model structures, the need to include multiple scattering in LEED calculations greatly increases the computational demands. This is particularly true if the size of the unit mesh of the surface structure is much larger than that of the underlying solid (typically more than about 5 times the area). For this reason many applications of SXRD to determine surface structures formed and studied under UHV conditions have focussed on systems in which the surface unit mesh is large.

In the early 1980s the 'holy grail' of surface structure determination was the (7 × 7) reconstruction of the Si(111) surface, when new models of the structure seemed to be proposed every few weeks in the pages of *Physical Review Letters*. In fact, the structural solution that has been widely accepted emerged from a technique not

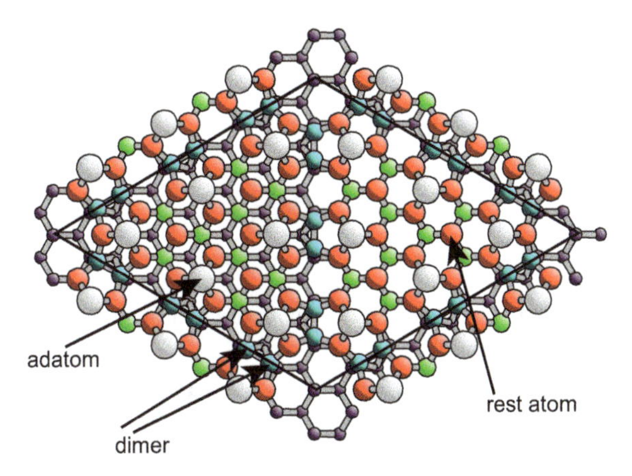

Figure 4.4. Schematic plan view of the DAS model of the Si(111)(7 × 7) surface reconstruction. The relative heights above the underlying solid are indicated by the radii of the atomic sphere, the lowest having the smallest radii, the highest the largest radii.

generally associated with surface structure determination, namely high-energy electron diffraction using an electron microscope, by Takayanagi *et al* (1985). This dimer-adatom-stacking fault (DAS) model is illustrated in figure 4.4. In bulk silicon (Si) all atoms are tetrahedrally bonded to four neighbouring atoms, so in an ideal bulk-terminated Si(111)(7 × 7) surface each of the 49 outermost Si atoms would have a single dangling bond perpendicular to the surface. Surface reconstruction occurs to reduce the number of these dangling bonds and thus lower the total surface energy. Within the unit mesh 12 Si adatoms occupy threefold coordinated sites above the outermost layer, thereby replacing 36 dangling bonds by just 12. A further reduction in the number of dangling bonds results from dimerisation of adjacent Si surface atoms along the edges of the unit mesh. Finally, a stacking fault is introduced into one half of the unit mesh, as seen in the difference on the two halves of the unit mesh in figure 4.4. The six surface Si atoms within the unit mesh that retain their dangling bonds are known as 'rest atoms'.

Achieving a complete quantitative structure determination of this structure, having ~100 Si atoms in displaced or additional sites relative to the ideally terminated bulk structure, is clearly a major challenge for any structural technique, but armed with the DAS model as an important starting point Robinson *et al* (1988) did show that SXRD could confirm and refine this structure, identifying local strain around the Si adatoms. Indeed, more recently Ciston *et al* (2009) have shown that using a combination of SXRD and high-energy electron diffraction it is possible to gain a detailed picture of the local charge density within this surface phase. Other examples of the application of SXRD to determine the structure of large unit mesh reconstructions on semiconductor surfaces include the c(8 × 2) phases of InSb, InAs and GaAs (Kumpf *et al* 2001).

SXRD has also been used successfully to determine the structure of adsorbates on surfaces including relatively large organic molecules. This latter application is challenging because these molecules contain mostly low-atomic-number atoms

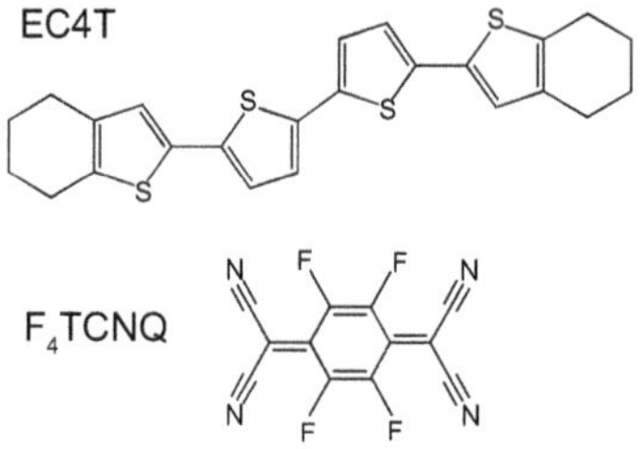

Figure 4.5. Molecular representations of EC4T and F_4TCNQ.

(mainly H, C, N, O), which are very weak X-ray scatterers. Despite this, the technique has been used in one example of end-capped quaterthiophine (EC4T; see figure 4.5) adsorbed on Ag(111) to show evidence of structural modifications in the molecular conformation caused by surface adsorption (Meyerheim *et al* 2000), an experiment actually achieved using a laboratory (18 kW rotating anode) X-ray source, albeit with very weakly scattered fractional-order diffraction peaks, with estimated errors in the intensities of 10%–25%. Although two CTRs were measured it did not prove possible to determine the local registry of the EC4T relative to the Ag(111) substrate. More recently, SXRD has been used to investigate the structure of F_4TCNQ (fully fluorinated 7,7,8,8-tetracyanoquinodimethane; see figure 4.5) on Au(111) (Mousley *et al* 2022). Based particularly on scanning tunnelling microscopy (STM) images, it had been proposed (Faraggi *et al* 2012) that in this adsorption system gold (Au) adatoms from the bulk are incorporated into the molecular overlayer to produce a two-dimensional metal–organic framework (or 2D-MOF), and indeed other studies of TCNQ adsorption had led to suggestions that this same effect may occur on other coinage metal surfaces. However, STM offers no means of identifying the atomic species giving rise to protrusions in these constant-tunnelling-current images, and indeed such protrusions cannot always be associated with atomic sites. The interpretation of such images can therefore be ambiguous. SXRD offers a method of providing direct evidence of the presence and location of high-atomic-number atoms within the molecular layer. The high X-ray scattering cross-sections for such atoms potentially dominates the diffracted intensities and is the basis of the 'heavy atom' method of solving complex macromolecular structures using standard XRD; it is also relevant to the related techniques of isomorphous replacement and multiwavelength anomalous dispersion (or MAD) XRD (e.g., Drenth 2007).

Like all XRD structural studies, SXRD suffers from the 'phase problem': a Fourier transform of the diffracted amplitudes could provide a direct picture of the real-space structure, but one can only measure intensities, not amplitudes, so the crucial phase information is lost. Nevertheless, a Fourier transform of the measured intensities, a Patterson function, can provide valuable clues to the real-space structure. In particular, if one performs a Fourier transform of the intensities of fractional-order-diffracted beams under 'in plane' conditions, i.e., nominally at zero values of l ($\Delta k_\perp$) (in practice at low values of l), the resulting Patterson map is a self-convolution of the real-space structure projected onto the surface plane. This is an image of all interatomic vectors in the surface structure, projected onto the surface, with amplitudes determined by the scattering cross-section of the participating atoms and the frequency of these vectors in the structure. Patterson maps have been used in many (especially early) SXRD studies of surface structures including the Ag(111)-EC4T system mentioned above, although other more sophisticated direct methods have also been developed (Marks *et al* 2001; see also the book chapter by Moritz and Van Hove 2022). Figure 4.6(a) shows such a Patterson map obtained from the experimentally measured 'in plane' fractional-order beams from the Au(111)-F_4TCNQ surface, while figure 4.6(b) shows the real-space structure

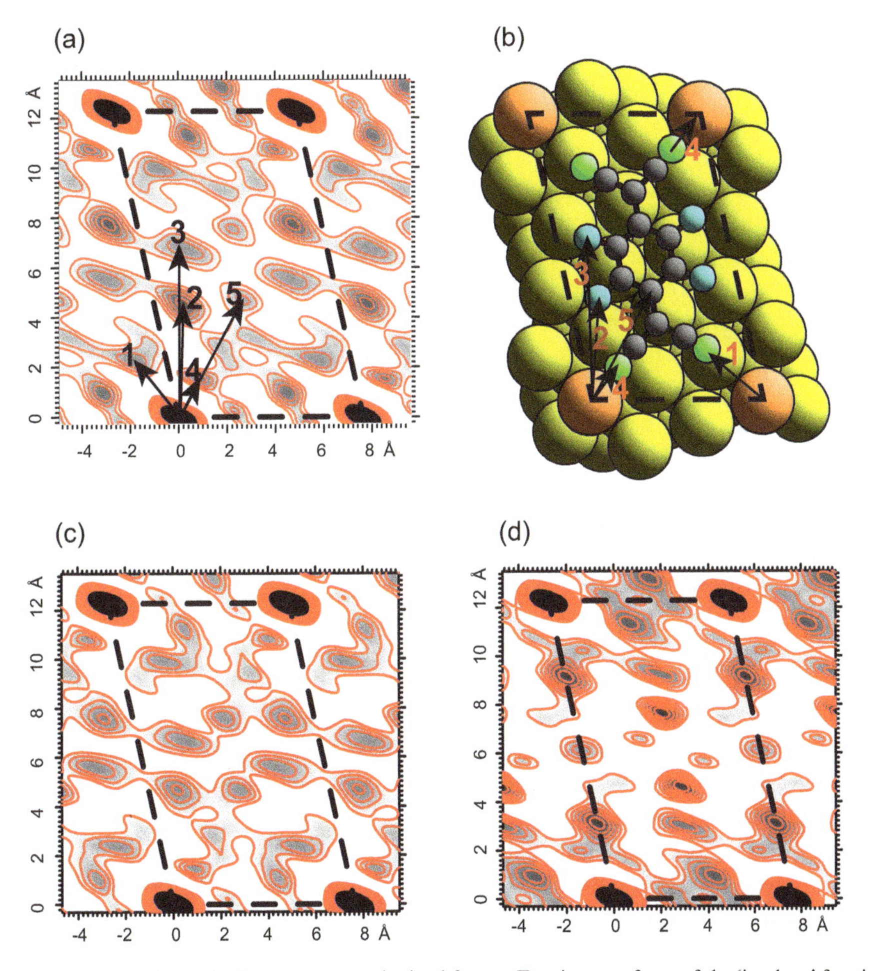

Figure 4.6. Panel (a) shows the Patterson map obtained from a Fourier transform of the 'in plane' fractional-order-diffracted beam intensities recorded from the Au(111)-F_4TCNQ surface. Panel (b) shows a representation of the predicted real-space structure by DFT calculations. Atom colouring: yellow, Au bulk; orange, Au adatoms; green, nitrogen; blue, fluorine. Superimposed on both panels are the equivalent set of interatomic vectors. Panels (c) and (d) show Patterson maps based on simulated in-plane diffracted intensities from the optimised DFT structure with (c) and without (d) Au adatoms. Panel (a) reprinted with permission from Mousley *et al* (2022). Copyright (2022) American Chemical Society.

according to the results of a dispersion-corrected density functional theory (DFT) calculation. In this representation of the structure the bulk Au atoms are shown as yellow spheres while Au adatoms are shown coloured orange. The colouring of the atoms in the molecule are as follows: nitrogen, green; carbon, grey; fluorine, blue. Superimposed on figure 4.6(a) are a series of numbered prominent interatomic vectors that may be derived from the map, while superimposed on the predicted real-space structure of figure 4.6 are the corresponding Au-N, Au-F and Au-C interatomic vectors. Notice that the strongest features in the experimental Patterson map define the size of the unit mesh; in a two-dimensionally periodic structure interatomic vectors between all equivalent atoms correspond to the primitive translation vectors of the unit mesh, so these are the dominant interatomic vectors. Features within the unit mesh in the Patterson function provide the structural information on the relative positions of atoms in the unit mesh and may be expected to be dominated by interatomic vectors between the strong Au scatterers and other atoms. The marked vectors are clearly consistent with the DFT-predicted structure. Further support for the essential correctness of the DFT mode structure (including Au adatoms) is provided by figures 4.6(c) and (d), which show Patterson maps constructed from simulated data based on the DFT model with adatoms and an optimised DFT structure without adatoms, respectively. Figure 4.6(c) is closely similar to the Patterson map generated by the experimental data (figure 4.6(a)), whereas figure 4.6(d) is quite different.

A more complete structure determination may be achieved by a trial-and-error iterative process, comparing experimental diffracted beam intensities with those predicted for series of trial structures using a computer program, most commonly the ANA-ROD package (Vlieg 2000), until the best fit is achieved. Moreover, while measured in-plane fractional-order beam intensities can provide the projection of the structure of the surface layers (those having the larger unit mesh than the underlying bulk), information on the relative heights of these atoms requires fitting of FOR scans; information on the surface–substrate registry is contained in the CTRs. This fuller analysis of the SXRD data from the Au(111)-F_4TCNQ phase did allow both the heights and the lateral registries of the F_4TCNQ molecules and Au adatoms relative to the underlying bulk to be obtained. However, the analysis proved to be insensitive to the exact conformation of the adsorbed molecules; complementary information on this aspect was obtained in separate experiments using the XSW technique described in section 4.3.

While these examples demonstrate the utility of SXRD for solving the structures of surfaces under UHV conditions, perhaps the most important area of application of the technique is to study surfaces and interfaces without the constraint of UHV, which applies to surface techniques that rely on the detection of scattered and emitted low-energy electrons. SXRD can therefore be used to investigate the structure of surfaces under ambient gases (including species associated with a surface chemical reaction) but also to study solid–liquid and solid–solid interfaces.

One example of this ability is in an investigation of the structure of rutile TiO_2(011) under a multilayer film of water (Hussain *et al* 2019). The interaction of water with TiO_2 surfaces has been subject to extensive study since it was shown that photochemical production of hydrogen from water can occur over titania (Fujishima and Honda 1972). Under standard UHV surface science conditions the interaction of water with the most-studied rutile $TiO_2(110)$ surface has led to the conclusion that H_2O dissociation only occurs at room temperature at surface defects (oxygen vacancies). There is evidence that dissociation on a stoichiometric surface does occur at low temperatures, but the resulting adsorbed hydrogen (H) and hydroxide (OH) species recombine and are desorbed as molecular water below room temperature (Duncan *et al* 2012). A range of structural studies (including by SXRD) have established that the (110) surface has some surface relaxations but retains the (1 × 1) periodicity of the underlying bulk. By contrast, similar UHV studies have shown that the clean (011) surface has a (2 × 1) reconstruction, but the quantitative structure when the surface is in contact with liquid water was not known, although a combination of STM and DFT calculations had led to the conclusion that the (2 × 1) reconstruction was lifted, leading to a (1 × 1) surface with coadsorbed H_2O and OH. The SXRD investigation, in a partial pressure of water of 30 mbar at room temperature, believed to lead to ~12 layers of molecular water on the surface, provided direct evidence of the transformation to a (1 × 1) structure. Of course, in this case there are no fractional-order-diffracted beams, so the surface structure determination relies entirely on CTR measurements. The conclusion of a quantitative analysis of these data is that the (1 × 1) surface is terminated with OH species at a Ti-O bond length consistent with DFT results.

While one of the main motivations for the development and application of modern surface science methods has been to understand key aspects of heterogeneous catalysis, most such studies have involved analysis of surfaces under static UHV conditions following exposure to reactive gases. There is certainly evidence that in some systems the reaction processes and mechanism are quite different at higher pressures, leading to the development of techniques that can 'bridge' the pressure gap. In general, photon in–photon out methods offer this possibility, and these certainly include SXRD. For example, exposure of many surfaces to oxygen gas can lead to chemisorbed oxygen phases that are then stable under UHV conditions, but additional oxidic phases can exist under higher oxygen pressures. As a specific example, consider the case of Pd(100). Under typical UHV surface science conditions oxygen (O) exposure leads to the formation of several chemisorption structures (2 × 2), c(2 × 2) (5 × 5) but also, at the highest exposures, a $(\sqrt{5} \times \sqrt{5})R27°$ structure that appears to comprise a single layer of palladium oxide (PdO; see below). Only under higher oxygen pressures does a true oxide phase form, and SXRD studies have revealed the structure of a multilayer epitaxial PdO under oxygen partial pressures in excess of 1 mbar and temperatures >650 K, but also a true bulk oxide under a pressure of 50 mbar at a temperature of 675 K (Stierle *et al* 2005).

Of course, in real heterogeneous catalysis the surface is in a steady-state dynamic equilibrium with much higher pressures of the reactive gases and products, and in recent years there has been increasing interest in extending these studies of model catalysis to more realistic reaction conditions. SXRD offers a way to investigate the surface structure under these conditions. One relatively simple but much-studied reaction is the oxidation of carbon monoxide (CO), particularly over palladium and platinum catalysts. The underlying process is clear: dissociative adsorption of O_2 on the surface makes available atomic oxygen that can interact with CO to produce CO_2, released into the gas phase. The reaction might occur between adsorbed O and adsorbed CO (the Langmuir–Hinshelwood mechanism) or between adsorbed O and impacting CO molecules from the gas phase (the Eley–Rideal mechanism), but a key structural question exists with regards to the active site. There is now significant evidence that a surface oxide is involved, and SXRD studies (complemented by other techniques including STM) have cast light on the nature of these phases, as reviewed by van Spronsen *et al* (2017).

Under reactions conditions three surface oxides have been identified on Pd(100), the $(\sqrt{5} \times \sqrt{5})R27°$ monolayer surface oxide, a multilayer (~2–3 nm thickness) epitaxial surface oxide and a bulk-like PdO that is no longer epitaxial. While SXRD offers the possibility of determining the details of such structures, collecting a complete data set of diffraction rods is potentially very time consuming, and is therefore difficult to obtain under steady-state reaction conditions. However, Gustafson *et al* (2014) showed that by performing the SXRD experiment at a much higher photon energy (85 keV) than the values of ~10–30 keV more commonly used, an extensive data set of rod intensities could be obtained much more rapidly. This high-energy SXRD (HESXRD) approach exploits the fact that as the photon energy is increased, the Ewald sphere radius is also increased, but also the angular range of the diffracted beam corresponding to a specific range of l $(\Delta k_\perp)$ associated with intersection of a reciprocal lattice rod is reduced (see figure 4.2). The grazing intersection of a near-planar part of the Ewald sphere with these rods means that a two-dimensional detector records intensity streaks that correspond to simultaneous measurements of whole sections of rod scans, greatly reducing the time required to obtain an adequate data set for structure determination (see also Nicklin 2014).

As remarked above, a $(\sqrt{5} \times \sqrt{5})R27°$ phase has been observed and studied in more conventional UHV surface science experiments of oxygen adsorption on Pd (100), although the formation of this phase requires high oxygen exposures and elevated temperatures. A number of quantitative LEED investigations have sought to determine the structure of this phase. Initially, the favoured structure was found to comprise a monolayer of PdO(001) in the surface (Saidy *et al* 2001), but a later study complemented by STM imaging and DFT calculations concluded that it comprises a strained layer of PdO(101) (Kostelník *et al* 2007). A quantitative HESXRD investigation of the structure of this phase under CO oxidation conditions at a total reactant gas pressure of 100 mbar, reported by Shipilin *et al* (2014), sought to distinguish between these basic models and a possible Pd(100)PdO(100) structure (see figure 4.7).

Comparisons of experimental rod scans with simulated data based on the three different structural models show good agreement for the Pd(100)PdO(101) and Pd (100)PdO(100) models but very poor agreement for the Pd(100)PdO(001) model; an example of two of these rod scans is shown in figure 4.8. However, the fits to the PdO

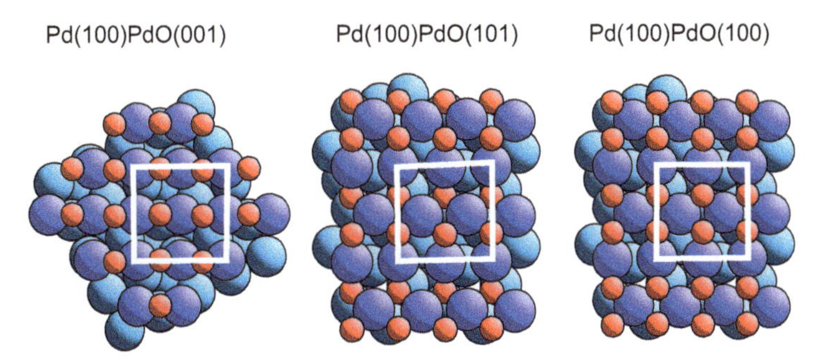

Figure 4.7. Plan views of alternative structural models for the $(\sqrt{5} \times \sqrt{5})R27°$ structural phase formed by oxygen on Pd(100) as considered in the SXRD study under CO oxidation conditions by Shipilin *et al* (2014). Substrate Pd atoms are shown pale blue, overlayer PdO Pd atoms are dark blue, and O atoms are red. The structural models of the Pd(100)PdO(001) and Pd(100)PdO(101) structures are based on the LEED studies by Saidy *et al* (2001) and Kostelnik *et al* (2007), respectively. The structural parameters of the Pd(100)PdO(100) model are those suggested by Shipilin *et al* (2014).

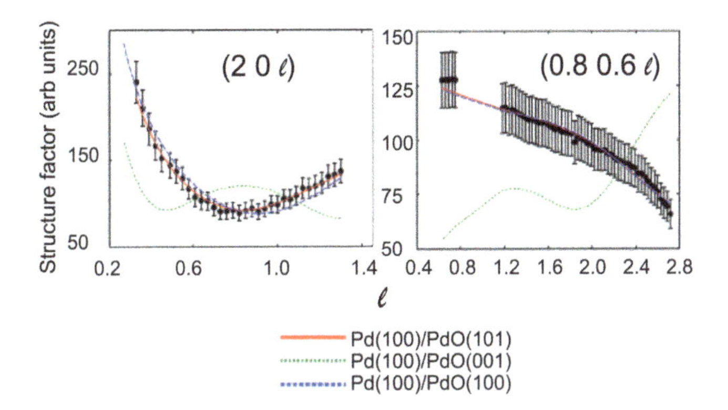

Figure 4.8. Comparison of two experimental rod scans from the SXRD study of Pd(100)-PdO by Shipilin *et al* (2014) with simulations for the three alternative structural model shown in figure 4.7. Reprinted from Shipilin *et al* (2014), copyright (2014), with permission from Elsevier.

(101) and PdO(100) models are essentially indistinguishable. The authors attribute this to the dominant role of the 'heavy' (high atomic number) Pd atom scattering in the oxide, relative to that from the O 'light' atoms. These two structural models differ only in that one half of the overlayer O atoms of the PdO(100) model are shifted below the outer Pd atom layer in the Pd(101) model. This is the same qualitative effect mentioned in the context of the Au(111)-F_4TCNQ surface described above. Indeed, this effect is also probably manifest in one of the very first applications of SXRD to an adsorbate-induced surface reconstruction, namely the Cu(110)(2 × 1)-O surface (Feidenhans'l *et al* 1990). This study clearly identified the location of the 'added rows' of Cu atoms in this phase, but the precision in the location of the O atoms was relatively poor, and indeed later studies by other methods led to rather different O atom heights above the surface.

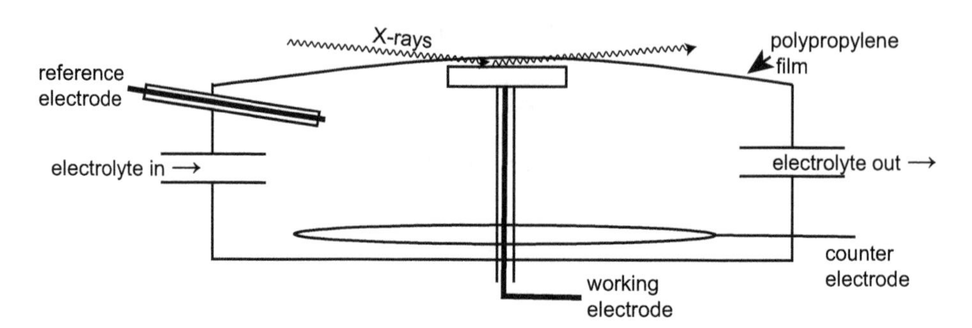

Figure 4.9. Simplified schematic diagram showing one design of an electrochemical cell allowing SXRD studies of a working electrode under a thin film of electrolyte.

The ability to investigate buried interfaces with SXRD has also been exploited to study electrochemical electrode–electrolyte interfaces, a special electrochemical cell such as that shown schematically in figure 4.9 allowing the incident and scattered X-rays to pass through a thin (~20 μm) film of electrolyte in front of a working electrode. The polypropylene film is porous, allowing external gas to be in equilibrium with the electrolyte. Notice that the cell contains three electrodes: the working electrode (the sample), and a reference and counter electrode. These are essential ingredients to perform cyclic voltammetry, the core characterisation tool of electrochemistry.

As in studies of gas–solid interactions many of the technological applications, such as corrosion and heterogeneous catalysis, involve polycrystalline surfaces or the surfaces of small particles, but studies of model systems based on single-crystal surfaces are also pursued in electrolyte–electrode interfaces, and there are some striking similarities in the interface behaviour. Thus, for example, the interaction of CO with the surface and CO oxidation reactions can be studied in both environments. Two different ordered phases of adsorbed CO have been identified (initially by *in situ* STM by Villegas and Weaver 1994) as present on Pt(111) in a CO-saturated electrolyte, a (2 × 2)-3CO phase with a CO coverage of 0.75 Ml and a $(\sqrt{19} \times \sqrt{19})R26.6°$-13CO phase with a CO coverage of 0.68 Ml (see figure 4.10). Figure 4.10 also shows how the transformation between these two phases occurs as the applied cell potentials are varied by monitoring the intensity of specific diffracted beams associated with the two different phases.

While there are many examples of close parallels between the observed structures of both clean surfaces and adsorbate-covered surfaces in gas adsorption and electrochemical environments, neither of these CO adsorption structures have been observed in UHV surface science experiments. Indeed, the highest CO coverage on Pt(111) in gas adsorption experiments appears to be 0.6 Ml (in the $c(\sqrt{3} \times 5)$rect.-CO phase). The possibility of achieving higher surface coverages at the electrochemical interface is consistent with the fact that varying the cell potential can produce a chemical potential that would be correspond to much higher CO pressures than can be studied in UHV experiments. The detailed structures of both phases have been investigated by SXRD, which shows local outward relaxation of Pt atoms

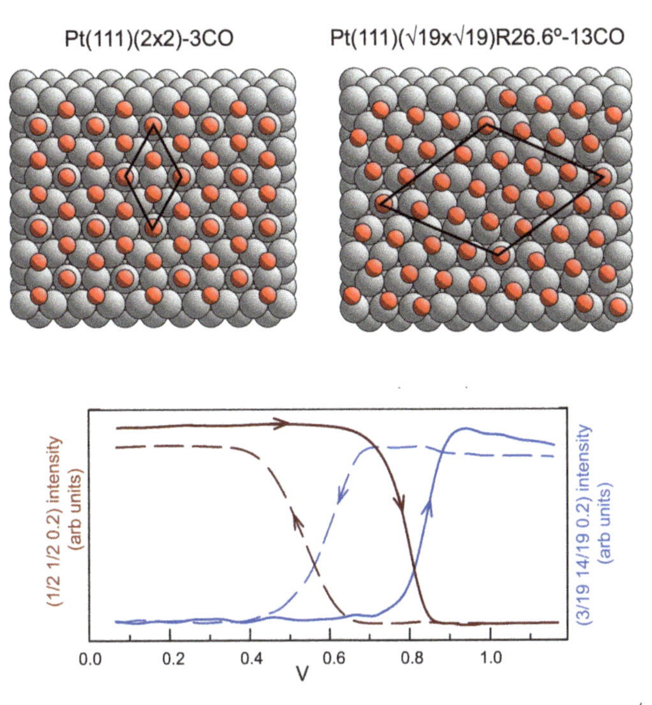

Figure 4.10. Schematic diagrams of the structure of the Pt(111)(2 × 2)-3CO and Pt(111)($\sqrt{19} \times \sqrt{19}$)R26.6°-13CO phases formed by CO adsorption at the electrochemical interface. Also shown below is the transformation between these two phases during cycling of the applied cell potential, monitored by the intensity of SXRD diffracted beams characteristic of these two phases. Reprinted from Lucas and Marković (2007), copyright (2007), with permission from Elsevier.

in the outermost layer. In the case of the ($\sqrt{19} \times \sqrt{19}$) phase, it was found that for the Pt atoms bonded to the CO molecules occupying atop sites this relaxation was 0.28 Å, whereas for the remaining Pt surface atoms to which CO is bonded in bridge and off-bridge sites, a much smaller relaxation of 0.04 Å was found (Wang *et al* 2005). In the (2 × 2) phase SXRD revealed a Pt outer-layer outward relaxation of 0.09 A (Lucas *et al* 1999). CO electrooxidation studies in acid solutions indicate that in the (2 × 2) phase the CO is rather weakly bonded to the surface, but a positive increase in the applied potential leads to an oxidative reduction in the local coverage to ~0.6 Ml, the remaining CO being more strongly bonded, which can then participate in the oxidation reaction at higher potentials (Marković *et al* 1999).

4.3 (Normal incidence) X-ray standing waves

A rather different technique for surface structure determination based on XRD exploits diffraction from the bulk substrate rather than the surface. When the direction and wavelength of X-rays incident on a crystalline solid meet the Bragg condition for the creation of a diffracted beam, the incident and diffracted beams interfere to produce an x-ray standing wave. This is shown schematically in figure 4.11(a). The periodicity of the intensity of this standing wave is equal to the

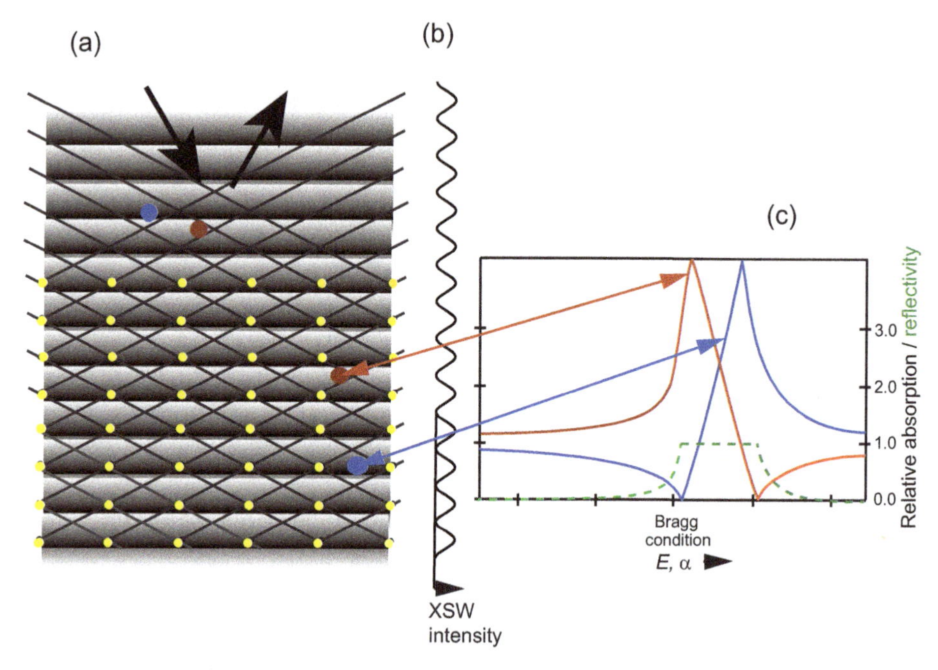

Figure 4.11. Panel (a) shows a schematic representation of the X-ray standing wave (XSW) created by the interference of an incident wave and the Bragg-diffracted wave from the planes of atoms (represented by the yellow circles) of a crystalline solid. Panel (b) shows the intensity variation of the standing wave as a function of distance perpendicular to the Bragg diffraction planes. Panel (c) shows the absorption profiles at the blue and red atoms in (a), assuming a non-absorbing crystal.

periodicity of the atomic spacing of the Bragg diffraction planes (figure 4.11(b)). To understand how this standing wave can be used to obtain structural information, it is important to recognise that a Bragg peak has a finite width, i.e., exists over a finite range of incidence angle or wavelength. In the simple derivation of Bragg's law one sums the scattering from all the atomic bases in each unit cell of an infinite perfectly three-dimensional periodic crystal. This treatment does give the Bragg condition as uniquely $2d \sin \theta = n\lambda$, where d is the spacing of the planes of atomic scatterers ('the Bragg planes'), θ is the grazing incidence angle to these planes, λ is the X-ray wavelength and n is an integer.

In reality, the scattering is only from scatterers within a finite depth, even from a semi-infinite crystal (there must be a surface), and this finite penetration depth leads to a finite range of θ or λ over which the standing wave can be formed. Notice that this penetration is finite even if there is no inelastic scattering. Of course, elastic scattering is a conservative process, but the presence of a surface means that X-rays scattered back out of the crystal cannot be scattered back into it. The back-scattering out of the crystal at each layer thus leads to a characteristic extinction depth. The significance of this unity reflectivity (for a non-absorbing crystal) is that the phase of the standing wavefield shifts, perpendicular to the Bragg planes, as one scans through this reflectivity range, by one half of the Bragg plane spacing. At one end of this range the antinodes of the standing wave are coincident with the Bragg planes, whereas at the other end of the range the nodal plans are coincident with the Bragg planes. Figure 4.11(a) shows the simple situation in which the atomic basis of the crystal structure comprises a single atom (such as in fcc or bcc solids) so the Bragg planes correspond to equivalent layers of these atoms. If the amplitude of the incident wave is A, then when the antinodes of the standing wave are coincident with the atomic planes the intensity of the radiation at these atoms is $(A + RA)^2$, and if the reflectivity $R = 1$, the intensity if $4A^2$. By contrast, when the nodes of the standing wave coincide with the atom planes the wavefield intensity at these atoms is $(A - A)^2 = 0$. In figure 4.11(a) the blue atom in the crystal is coincident with these Bragg planes, leading to the blue profile (calculated for a non-absorbing crystal) shown in figure 4.11(c). If one monitors the X-ray absorption at these atoms as one scans through the Bragg condition there is a characteristic absorption profile determined by the location of the absorbers relative to the Bragg planes. If one has absorbers midway between the Bragg planes (such as the red atom in figure 4.11(a)), the intensities are reversed as shown by the red curve in figure 4.11(c). Notice that the incident and diffracted beams overlap outside the crystal surface, so atoms on the surface are embedded in the standing wavefield in the same way as atoms within the crystal.

If one measures the X-ray adsorption at specific atoms in, or on, the crystal as one scans the incident angle or energy, one therefore obtains an absorption profile characteristic of the location of these atoms relative to the Bragg planes. Figure 4.12 illustrates this effect with calculations based on the (111) Bragg condition measured with normal incidence to a Cu(111) sample, showing the variation in absorption as a function of photon energy (and hence wavelength) for absorbed atoms at heights above the (111) atom planes of 0, 0.5, 1.0 and 1.5 Å. Notice that the Bragg plane spacing in this case is 2.08 Å, so these intermediate heights correspond to approximately 25%, 50% and 75% of the Bragg plane spacing. The very pronounced difference in these absorption profiles clearly indicates that the technique can measure these layer heights with sub-ångström precision. Notice that these

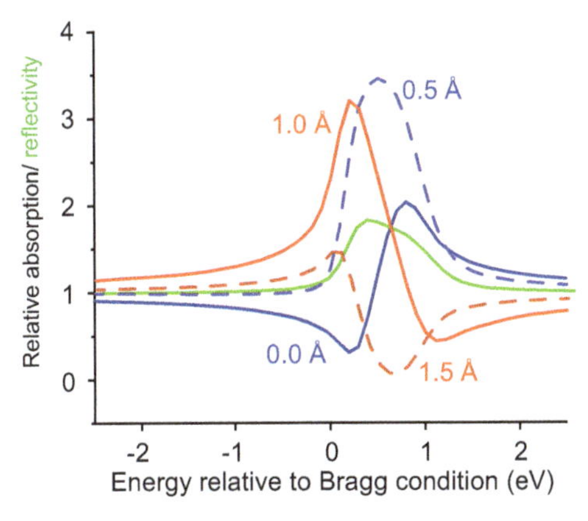

Figure 4.12. Calculated XSW absorption profiles for absorbers at different heights above the atomic planes of a Cu(111) sample around the normal incidence (111) Bragg condition. Also shown is the reflectivity.

calculations are for a real material (copper) including the effects of absorption, which leads to some weakening of the amplitude of the modulations and a distortion of the reflectivity profile. In order to measure these absorption profiles experimentally in an element-specific fashion one can detect the X-ray fluorescence or Auger electron emission associated with the refilling of the core holes following absorption. As discussed below, one can also use detection of photoelectrons emitted as a direct result of the absorption, although this leads to profiles that differ slightly from a true absorption measurement. Before discussing this aspect we need to provide a quantitative description of the underlying physics.

The XSW intensity can be written as

$$I = \left| 1 + \frac{E_H}{E_0} \exp(-2\pi i z / d_H) \right|^2, \tag{4.1}$$

where E_H and E_0 are the amplitudes of the electromagnetic field of the scattered and incident X-rays (H represents the (hkl) Bragg planes), d_H is the spacing of the Bragg planes and z is the height above one of these planes. The amplitude ratio is given by

$$\frac{E_H}{E_0} = -\left(\frac{F_H}{F_{\overline{H}}}\right)^{1/2} [\eta \pm (\eta^2 - 1)^{1/2}], \tag{4.2}$$

where F_H and $F_{\overline{H}}$ are the structure factors for (hkl) and $(-h-k-l)$, respectively. η is a displacement parameter determined by the deviation of incidence angle, $\Delta\theta$, or photon energy, ΔE, from the exact Bragg condition. In the case in which one scans θ through the Bragg peak (commonly referred to as a 'rocking curve'),

$$\eta = \frac{(\Delta\theta \sin(2\theta_B) + \Gamma F_0)}{|P|\Gamma(F_H F_{\overline{H}})^{1/2}},$$

where θ_B is the nominal Bragg angle, P is a polarisation factor ($P = 1$ for σ polarisation and $P = \cos(2\theta_B)$ for π polarisation), F_0 is the structure factor for forward-scattering ($(hkl) = (000)$), and

$$\Gamma = \frac{e^2}{4\pi\varepsilon_0 mc^2} \frac{\lambda^2}{\pi V},$$

where e and m are the electron charge and mass, ε_0 the permittivity of free space and V the volume of the unit cell. In the case of a non-absorbing crystal, for which F_H and $F_{\overline{H}}$ are both real, equation (4.2) shows that the reflectivity is unity for the range $1 \geqslant \eta \geqslant -1$. Notice that η is not zero when $\Delta\theta = 0$, so the unity reflectivity range and associated absorption profiles are not centred on the exact Bragg condition, as is evident in figures 4.11(c) and 4.12. This offset is due to the effect of multiple forward-scattering of the X-rays, which accounts for the attenuation due to elastic back-scattering. The range of $\Delta\theta$ corresponding to the range $1 \geqslant \eta \geqslant -1$ is given by

$$\text{range}\,(\Delta\theta) = \pm[|P|\Gamma(F_H F_{\overline{H}})^{1/2} / \sin(2\theta_B)].$$

A numerical evaluation of this range for a specific material leads to an important conclusion regarding the application of the XSW technique to surface adsorption structures. Considering again the example of a (111) Bragg condition in Cu, and taking the wavelength to be 1 Å (a photon energy of 12.4 keV) corresponds to a value of θ_B of 13.9°, which leads to the total range of $\Delta\theta$ leading to the standing wave being approximately 15 µrad or 3 arc seconds. This 'rocking curve' width is clearly very small. The implication of this result is that a viable XSW experiment from Cu(111) would require incident radiation that is extremely highly collimated, but also that the Cu crystal would have to have an exceptionally high degree of perfection. Typical metal single crystals used in surface science studies may have a mosaicity (i.e., a range of different orientations due to small angle grain boundaries) of order 0.1° or more, which would smear out the absorption profile so much as to be barely detectable. This need for highly perfect crystals led for several years to the XSW technique being applied only to silicon (or germanium) crystals that can be grown to exceptionally high degrees of perfection (indeed, they are often described as 'dislocation free'). However, this problem can be circumvented, and XSW can be performed on standard metal single crystals by performing the experiment with a Bragg angle close to 90° (i.e., with normal incidence to the Bragg planes). At this angle the Bragg condition ($2d\sin\theta = n\lambda$) passes through a turning point (the gradient is proportional to $d(\sin\theta)/d\theta = \cos\theta$ so at $\theta = 90°$ the gradient is formally 0), so the Bragg condition has very low sensitivity to the exact value of θ. Modest collimation and mosaicity of order 0.1° is thus acceptable. This is the origin of the use of normal incidence X-ray standing waves (NIXSW). Notice that the requirement of normal incidence is relative to the Bragg planes, although, in practice, for many experiments one chooses Bragg planes parallel to the surface, so one also uses normal incidence to the surface in order to determine the height of absorber atoms relative to the surface. Using a fixed angle of incidence of 90° one can then scan through the Bragg condition by varying the photon energy (and thus the X-ray wavelength). The displacement parameter η can then be expressed in terms of the energy displacement, ΔE, from the exact Bragg condition as

$$\eta = \frac{[-2(\Delta E/E)\sin^2\theta_B + \Gamma F_0]}{|P|\Gamma(F_H F_{\overline{H}})^{1/2}},$$

leading to an energy range over which the standing wave exists of

$$\text{range}(\Delta E) = \pm[E|P|(F_H F_{\overline{H}})^{1/2}/2\sin^2 2\theta_B].$$

While the use of the normal incidence condition in SXW experiments offers the considerable advantage of making the technique applicable to a very wide range of materials that cannot, in general, be obtained as exceptionally perfect single crystals, it does constrain the X-ray energy that is required to a value consistent with satisfying the Bragg condition at normal incidence, i.e., at a photon energy equal to $2d/n$. This has some implications for the method of measuring the X-ray absorption at specific atoms, as will be discussed below. Notice, too, that for values of d of ~2–3 Å, as in many elemental crystals, the required wavelengths of ~4–6 Å correspond to photon energies of ~2–3 keV, an energy range significantly lower than that of 'hard' X-rays (of ~8 keV or more) used in a wide range of XRD experiments at synchrotron radiation facilities.

A measurement of the variation in absorption at a specific atomic species as one scans through the Bragg condition provides a measure of the relative intensity of the standing wave as given by equation (4.1), but this equation assumes that all the absorbers being studied have identical heights, z, above the Bragg planes. In practice, there will be a range of heights occupied, if only because of thermal vibrations. To analyse this situation, it is helpful to recast equation (4.1) in terms of the reflectivity, R. $R = (E_H/E_0)^2$, so one can write $E_H/E_0 = \sqrt{R}\,\exp(i\Phi)$ and thus $I = |1 + \sqrt{R}\,\exp(i\Phi - (2\pi iz/d_H))|^2$, which expands to become

$$I = 1 + R + 2\sqrt{R}\,\cos(\Phi - (2\pi z/d_H)). \tag{4.3}$$

If we now assume that the absorbers occupy a distribution of heights defined by $f(z)$ normalised such that $\int_0^{d_H} f(z) = 1$, then equation (4.3) becomes

$$I = 1 + R + 2\sqrt{R}\int_0^{d_H} f(z)\cos(\Phi - (2\pi z/d_H))dz,$$

which can be written as

$$I = 1 + R + 2f_{co}\sqrt{R}\,\cos(\Phi - 2\pi p). \tag{4.4}$$

This equation defines two key parameters, f_{co}, the *coherent fraction*, and p, the *coherent position*. These two parameters completely determine a measured absorption profile. p is a distance in units of the Bragg plane spacing d_H ($p = D/d_H$, with D a physical height), while f_{co} is an order parameter (if $f_{co} = 1$, equation (4.4) reduces to equation (4.3) that defines the absorption profile for an absorber at a single height z). The influence of f_{co} as an order parameter becomes clear if one rewrites equation (4.4) as

$$I = f_{co}(1 + R + 2\sqrt{R}\,\cos(\Phi - 2\pi p)) + (1 - f_{co})(1 + R).$$

The first term on the right-hand side of this equation is the coherent part of the absorption in the standing wave for a single absorber height as in equation (4.3) (with $z = pd_h$), attenuated by the factor f_{co}, while the second term is the incoherent part of the absorption (with a prefactor $(1 - f_{co})$) due to the background which is not part of the standing wave. Providing that f_{co} is found to have a value reasonably close to unity (this vague statement will be quantified below), the height of the absorbers above the nearest extended Bragg plane is given by $D = (p + m)d_H$, where m is an integer. Notice that the Bragg planes are periodic and the *extended* Bragg planes continue to repeat above the physical surface. The ambiguity in the unknown value of m can almost always be resolved rather easily by ensuring the implied interatomic distances are physically reasonable.

As remarked above, a particularly common application of the NIXSW technique is to determine the height of atoms and molecules above a surface using normal

X-ray incidence relative to the surface (using Bragg planes parallel to the surface). In many studies of the adsorption of essentially planar π-bonded molecules this information alone can prove to be very valuable in understanding the bonding to the underlying surface. A more complete structure determination, however, also requires knowledge of the lateral registry of the adsorbate relative to the surface. This can be achieved by simple triangulation taking NIXSW measurements relative to Bragg planes that are *not* parallel to the surface. A particularly simple case arises in studies of adsorption on fcc (111) surfaces, NIXSW measurements using the (111) Bragg planes giving the height of the absorber while similar measurements (in this case using the same photon energy range) of NIXSWs using a $\{\bar{1}11\}$ Bragg plane provide information on the lateral registry (see figure 4.13). Of course, in general one requires at least three measurements in different directions to provide complete triangulation, but because of the threefold rotational symmetry of this surface two measurements are sufficient.

In particular, figure 4.13 shows how adsorption in each of the three highest-symmetry adsorption sites can be distinguished. This shows that these three sites correspond to the adsorbate being atop a top-layer substrate atom ('atop'), atop a second-layer substrate atom ('hcp hollow') and atop a third-layer substrate atom ('fcc hollow'). The angle between the (111) and $(\bar{1}11)$ Bragg planes is 70.4°, so if the adsorbate occupies an atop site the $(\bar{1}11)$ NIXSW experiment will yield a $D_{(\bar{1}11)}$ value of $D_{(111)}\cos(70.4) = D_{(111)}/3$. By contrast, the values of $D_{(\bar{1}11)}$ for the hcp and fcc hollows are, respectively, $(D_{(111)}+d_{(111)})/3$ and $(D_{(111)}+2d_{(111)})/3$, which are easily distinguished. The results of an early application of this with interesting consequences arose in a NIXSW comparative study of the adsorption of atomic oxygen and of the de-protonated methanol species, methoxy (CH_3O-) on Al(111). Figure 4.14 shows the NIXSW absorption profiles at the (111) and $(\bar{1}11)$ Bragg conditions obtained at the Al and O atomic sites, together with theoretical fits to each

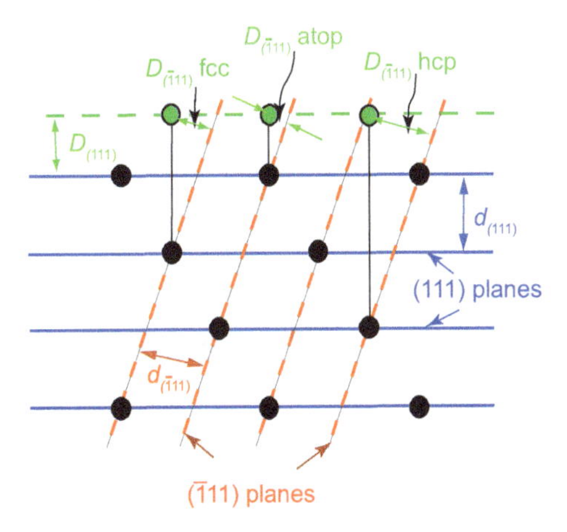

Figure 4.13. Side view of the (111) surface of a fcc solid with the atoms represented by black circles, showing the (111) (blue) and $(\bar{1}11)$ (red) Bragg planes. Adsorbate atoms represented by green circles are shown in atop, fcc hollows and hcp hollows at the same height, $D_{(111)}$. Also shown are the $D_{(\bar{1}11)}$ values obtained by $(\bar{1}11)$ NIXSW measurements for each of the three high-symmetry adsorption sites.

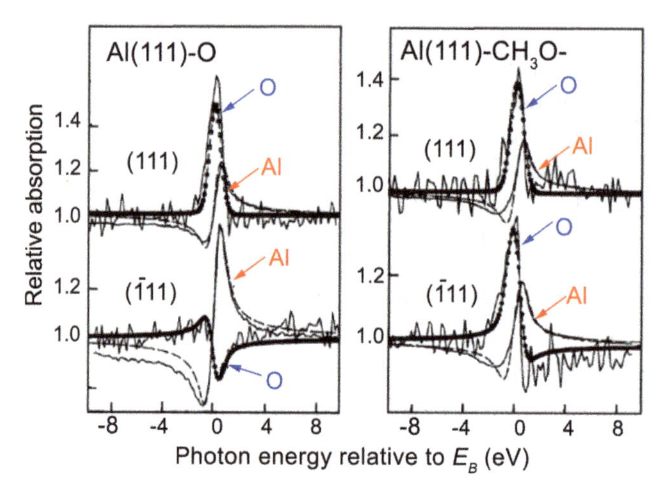

Figure 4.14. NIXSW absorption profiles measured at Al and O atoms for chemisorbed O and methoxy on Al (111), recorded around the (111) and ($\bar{1}$11) Bragg conditions. Smooth lines are the theoretical results giving the best fit to the noisier experimental data. Reproduced from Kerkar *et al* (1992).

experimental absorption profile. The (111) data are essentially identical for the two different adsorbates and are fitted by coherent position values corresponding to a value of the height of the O atoms above the outermost aluminium (Al) layer, $D_{(111)}$ = 0.7 ± 0.10 Å. However, the O absorption profiles at the ($\bar{1}$11) normal incidence Bragg condition are quite different. Specifically, for atomic O $D_{(\bar{1}11)}$ = 1.75 Å, whereas for the O atoms in the adsorbed methoxy species $D_{(\bar{1}11)}$ = 0.95 Å. For Al, $d_{(111)}$ = 2.33 Å, so for a value of $D_{(111)}$ of 0.7 ± 0.10 Å the expected values of $D_{(\bar{1}11)}$ for O atoms in the atop, hcp hollow and fcc hollow sites are, respectively, 0.23 ± 0.04 Å, 1.79 ± 0.04 Å and 1.01 ± 0.04 A. Evidently, the experimental value of $D_{(\bar{1}11)}$ for the chemisorbed atomic oxygen is consistent with adsorption in fcc hollow sites (a conclusion consistent with the structure determination of this surface by other methods). By contrast, the experimental value of $D_{(\bar{1}11)}$ for the adsorbed methoxy species is consistent with the O atom occupying the hcp hollow site. The occupation of this rather unusual adsorption site is discussed by Kerkar *et al* (1992).

Notice that in this discussion $D_{(111)}$ has been equated with the height of the O absorbers above the outermost Al (111) atomic layer. Strictly, this is not correct. NIXSW measures the location of an absorber relative to the extended Bragg planes of the underlying bulk crystal. These will be coincident with the outermost atomic Al planes only if the crystal is bulk terminated, with no relaxation of the outermost Al interlayer spacings. In practice, fcc (111) metal surfaces tend to show only very small near-surface layer relaxations, so the assumption of ideal bulk termination is unlikely to lead to a significant error in the implied local structure in this particular case, but the influence of surface relaxation needs to be borne in mind in interpreting such data.

The discussion of this case has considered only the heights of absorber atoms above the relevant Bragg planes, assuming that these heights can be obtained directly from the measured coherent positions. As such, the experimentally determined values of the coherent fractions have not been considered, although, importantly, these values were found to be ~0.9, quite close to the maximum possible value of unity. The significance of this, and the importance of the value of the coherent fraction in general, can be appreciated by noting that the coherent fraction f_{co} and the coherent position p can be related to the actual distribution of absorber heights, $f(z)$, by

$$f_{co} e^{2\pi i p} = \int_0^d f(z) e^{2\pi i (z/d)} dz.$$

This equation can be rather conveniently represented on an Argand diagram; the left-hand side is represented by a vector of length f_{co} and direction determined by the angle $2\pi p$ (figure 4.15(a)), while the right-hand side is represented by a range of vectors of length $f(z)$ and angle $2\pi z/d$.

Figures 4.15(b), (c) and (d) show representations of the coherent fractions and positions arising from equal co-occupation of two different values of z. In the case of figure 4.15(b), the two values of z, z_1 and z_2, differ by less than $d/2$, in which case the resultant height, pd, is simply the *average* of the two contributing z values, i.e., $p = (z_1 + z_2)/2d$. The resultant coherent fraction falls as the difference in the contributing heights increases, $f_{co} = f_0 |\cos(\pi(z_1 - z_2)/d)|$, where f_0 specifies the fractional occupation of the sites. In figure 4.15(c) is shown the case in which the difference in the two contributing z values, Δz, is given by $d > \Delta z > d/2$. In this case, the implied height pd is not the average of the two component heights. Figure 4.15(d) shows the special case in which $\Delta z = d/2$. The two component vectors are now diametrically opposed such that $f_{co} = 0$ and p is undefined. This case highlights the fact that f_{co} is not simply an 'order parameter' as is often implied. Here is a case of potentially perfect order (albeit with co-occupation of two heights), yet with a coherent fraction of zero.

Notice that for this case of two equally co-occupied heights f_{co} is a periodic function of Δz, because NIXSWs are only sensitive to the absorbed height relative to the nearest Bragg plane below the absorber, while Bragg planes are periodic. Figure 4.16 shows how the coherent fraction varies as a function of Δz.

Evidently, if the determined coherent fraction is high enough (close to unity), the coherent position can be directly related to the actual (singly) occupied height; but how high is 'high enough', and what can lead to a value of less than unity in a real experiment? The most obvious, and indeed inevitable, source of 'disorder' is the effect of thermal vibrations. Harmonic thermal vibrations lead to a Gaussian distribution of absorber heights, the resultant coherent position corresponding to the average value of the height distribution, while the coherent fraction is reduced by an amount that depends on the amplitude of the vibrations. In fact, atomic vibrations lead to a reduction in the coherent fraction for two reasons. Firstly, vibrations of the atoms in the underlying bulk crystal lead to a reduction in the amplitude of the standing wavefield, by a Debye–Waller factor e^{-2M}, the 'missing' amplitude appearing as thermal diffuse scattering. For the case of a Bragg angle of 90° as in a NIXSW experiment,

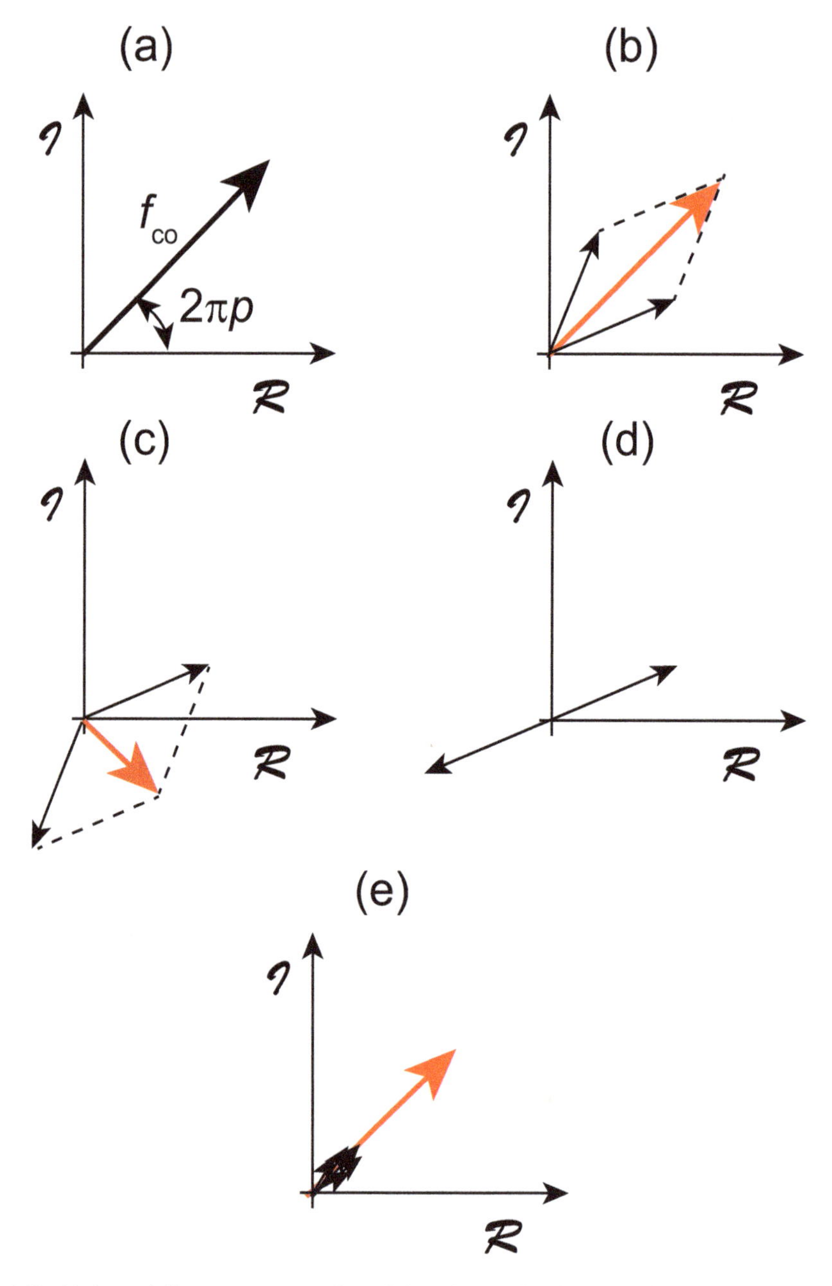

Figure 4.15. (a) Argand diagram representation of the coherent fraction f_{co} and coherent position p of the results of fitting NIXSW data. Panels (b), (c) and (d) show representations of the sum of two equally occupied component z values (in black) and the resultant sum (in red). Panel (e) shows a simplified representation of the sum of a narrow distribution of different z values such as arises from thermal vibrations of the absorber location. Reproduced from Woodruff and Duncan 2020. © IOP Publishing Ltd. CC BY 4.0.

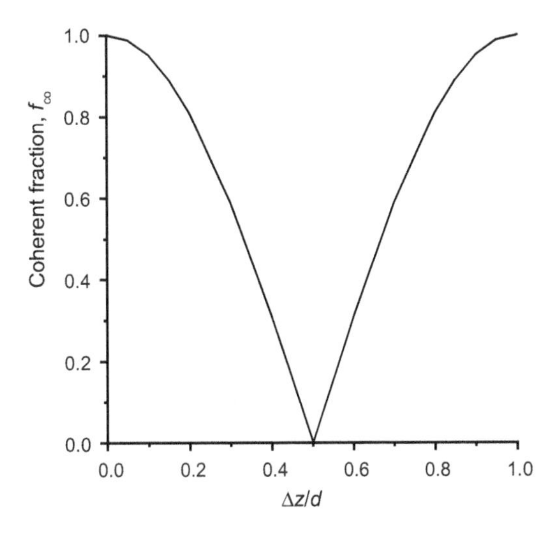

Figure 4.16. Variation of coherent fraction for the sum of two equally occupied heights as a function of the height difference, Δz. Reproduced from Woodruff and Duncan (2020). © IOP Publishing Ltd. CC BY 4.0.

$$M = \frac{8\pi^2}{\lambda^2}\langle u^2\rangle,$$

where $\langle u^2\rangle$ is the mean-square vibrational amplitude of the substrate atoms. In practice, at room temperature and for a typical elemental metal substrate, this reduces f_{co} by about 5%. A further reduction of the coherent fraction arises from the displacements of the absorber atom as illustrated schematically in figure 4.15(e). This reduction affects the amplitude rather than the intensity of the standing wave, so the attenuation is by a factor e^{-M}, where in this case $\langle u^2\rangle$ is the mean-square displacement of the absorber atoms due to both atomic vibrations and static disorder. In some NIXSW studies there has been a tendency to assume that low values of the coherent fraction are due to 'disorder' of part of the material under study, implying that some fraction of the sample is 'totally disordered' (i.e., $f_{co} = 0$) while the remaining part is well-ordered such that the recorded coherent position corresponds to the true atomic heights in this ordered part. However, for the most commonly studied systems comprising molecular overlayers with only a single molecule thickness, and with NIXSW being performed with normal incidence to the surface such that the coherent position corresponds to a layer spacing, consideration of the possible types of disorder in such systems (see Woodruff and Duncan 2020) shows clearly that 'total disorder' *perpendicular to the surface* cannot occur, and that disorder in such systems cannot account for coherent fractions much less than ~0.7. An important conclusion is that lower coherent fractions imply that at least two distinctly different atomic heights must be co-occupied and that as such the recorded coherent position does not correspond to the true atomic height of the absorber atoms. In such systems one type of 'disorder' that can account for low coherent fractions is the presence of more than one molecular layer. Indeed, a thick multilayer polycrystalline film *can* lead to a coherent fraction of zero for that part of a surface.

Notice that the key measurement in a NIXSW experiment is the variation in X-ray absorption at the atomic species of interest as the photon energy is scanned through the Bragg condition. In practice, for a sub-monolayer coverage of an adsorbate on the surface of a single crystal with a thickness of the order of millimetres, direct measurement of absorption by the reduction in transmittance through this sample evidently lacks the sensitivity to detect absorption in the overlayer. The measurement of the absorption must therefore be performed indirectly. The element-specific physical mechanism of X-ray absorption is photoionisation, leading to excitation of an electron in a core level into the continuum (photoemission), so absorption can be detected either by detecting the emitted photoelectron or by monitoring the refilling of the resultant core hole that occurs by X-ray fluorescence or Auger electron emission. Of these latter two modes of detection only Auger electron emission offers surface specificity due to inelastic scattering of the emitted electrons; fluorescent X-ray emission offers a way of detecting X-ray absorption at atoms within the crystal, but can only provide surface-specific information if the emitting elements are known to only be present in the surface. The main disadvantage with Auger electron detection is that Auger electron peaks sit on a high background of secondary electron emission, a problem exacerbated by the fact that the most intense Auger peaks are generally from core–valence–valence transitions, leading to broad peaks that are more difficult to separate from this high background. By contrast, the peaks in the emitted electron energy spectrum arising from core-level photoemission are generally narrow and thus much more easily separated from the background. Moreover, while the energy of emitted Auger electrons and photoelectrons are element specific, small 'chemical' shifts in core-level photoemission peaks (as discussed in chapter 3) are readily detected, so photoelectron detection offers not only element specificity but also chemical-state specificity.

There is, however, a complication in the use of photoemission detection. While the *total* photoemission (integrated over 4π steradians) is proportional to the total photoabsorption, energy-selective photoelectron detection using a dispersive analyser necessarily detects only photoemission in a limited angular range, and the intrinsic angular dependence of the photoemission means that the detected angle-resolved signal is no longer directly proportional to the total photoabsorption. The angular dependence of photoemission from an atom is determined by the matrix element between the initial core-level state, i, and the outgoing photoelectron final state, f:

$$M_{\rm fi} = \langle f|\exp(2\pi i\mathbf{k}.\,\mathbf{r})\mathbf{A}.\,\mathbf{p}|i\rangle,$$

where $\mathbf{k}$ is the photon wavevector, $\mathbf{r}$ is the electron position vector, $\mathbf{A}$ is the photon polarisation vector and $\mathbf{p}$ is the electron momentum operator. The exponential can be expanded as

$$\exp(2\pi i\mathbf{k}.\,\mathbf{r}) = 1 + 2\pi i\mathbf{k}.\,\mathbf{r} + \mathrm{O}(2\pi i\mathbf{k}.\,\mathbf{r})^2 + \cdots. \tag{4.5}$$

Generally, it is assumed that the dipole approximation is valid so that only the first term on the right-hand side of this equation is assumed to be non-zero, i.e., that $\exp(2\pi i\mathbf{k}.\,\mathbf{r}) \approx 1$. The underlying assumption is that the spatial variation of the electromagnetic wavefield over the initial-state wavefunction is negligible. In this case, the angular dependence of the photoemission has the form

$$\frac{d\sigma}{d\Omega} = \left(\frac{\sigma}{4\pi}\right)\left[1 + \left(\frac{\beta}{2}\right)\left(3\cos^2\theta_p - 1\right)\right],$$

where θ_p is the angle between the photoelectron detection direction and the polarisation vector $\mathbf{A}$. β is the dipolar asymmetry factor, which can take values between −1 and 2 for different specific initial states. The simplest case is emission from an initial *s*-state (orbital quantum number $l = 0$), when the dipole selection rule $\Delta l = \pm 1$ only allows one outgoing state, namely a *p*-wave ($l = 1$), leading to a value of $\beta = 2$ such that $d\sigma/d\Omega \propto \cos^2\theta_p$. Figure 4.17 shows this angular dependence arising from an incident photon wavevector $\mathbf{k}_0$ and a reflected (diffracted) photon wavevector $\mathbf{k}_H$ at an arbitrary angle of incidence to a sample. Clearly the angular dependence of the photoemission is such that the detected photoemission signal from the incident and reflected waves differ. Of course, if there is coherent interference between the two waves to produce a standing wave the true detected photoemission signal will differ from the incoherent sum of these two excitations, but the fact that the detected photoemission signal will not be truly proportional to the total photoionisation cross-section remains correct. However, for the case of normal incidence to the sample, $\mathbf{A}_0$ and $\mathbf{A}_H$ become colinear so the angle θ_p is the same for the incident and reflected waves ($\theta_{p0} = \theta_{pH}$) and the detected photoemission *is* proportional to the total photoionisation cross-section. NIXSW detection of (angle-resolved) photoemission from an initial *s*-state is therefore a true monitor of the photoionisation cross-section if the dipole approximation is valid.

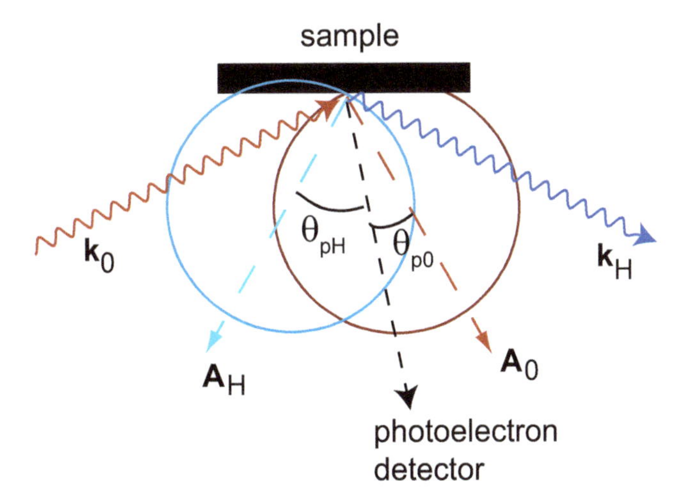

Figure 4.17. Schematic diagram showing the angular dependence of photoemission from an initial *s*-state arising from an incident photon wavevector $\mathbf{k}_0$ and a diffracted photon wavevector $\mathbf{k}_H$.

The validity of the dipole approximation is widely assumed to be appropriate in a wide range of photoemission experiments, conventional wisdom for many years having been that non-dipolar effects are only important at high photon energies (greater than or of the order of 20 keV). However, NIXSW experiments with photon energies of as low as ~3 keV have shown that non-dipolar effects can be significant. The important conclusion is that non-dipolar effects lead to a 'backwards–forwards' asymmetry in the angular dependence of the photoemission, dependent on whether the photoemission direction contains a positive or negative component relative to the exciting photon propagation direction. Figure 4.18 illustrates this schematically for the case of normal incidence, but for an atomic photoemission angular dependence that displays backwards–forwards asymmetry. Clearly this asymmetry leads to the (incoherent) photoemission from the incident and reflected X-rays being sampled differently.

As described above, the dipole approximation corresponds to taking only the first term (unity) from the right-hand side of equation (4.5). The next approximation is to take the first two terms, so that one assumes that $\exp(2\pi i\mathbf{k}.\,\mathbf{r}) = 1 + 2\pi i\mathbf{k}.\,\mathbf{r}$, which leads to two new terms in the matrix element describing the electric quadrupole (E2) and magnetic dipole (M1) interactions in addition to the electric dipole (E1) term. The photoemission probability depends on the modulus squared of the matrix element, $(E1+E2+M1)^2$, which is dominated by the pure dipole term $E1^2$, but while $E2^2$, $M1^2$ and E2M1 are very small, the cross-terms E1E2 and E1M1 are not negligible and lead to new angular effects. Specifically, including these terms leads to an angular dependence:

$$\frac{d\sigma}{d\Omega} = \left(\frac{\sigma}{4\pi}\right)\left[1 + \left(\frac{\beta}{2}\right)\left(3\cos^2\theta_p - 1\right) + \left(\delta + \gamma\cos^2\theta_p\right)\sin\theta_p\cos\phi\right],$$

where δ and γ are the additional asymmetry parameters arising from the E1M1 and E1E2 interference terms, respectively, while ϕ is the angle between the direction of photon propagation and the projection of the photoelectron wavevector in the plane

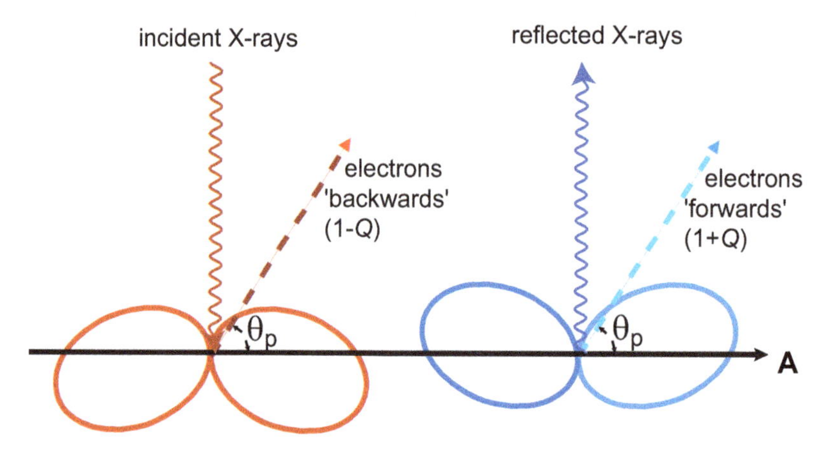

Figure 4.18. Schematic diagram showing the effect of backwards–forwards asymmetry in angular dependence of photoemission from incident and reflected X-rays in NIXSW.

perpendicular to **A**. $\cos\phi$ therefore takes a value of +1 or −1 depending on whether the photoemission is detected in the forwards or backwards direction relative to the direction of the photon propagation. This gives rise to the backwards–forwards asymmetry described above, which can be quantified by a backward–forward asymmetry parameter Q, such that the ratio of the intensity in the opposite directions for the same value of θ_p is (1+Q)/(1−Q). Evidently, in terms of δ and γ,

$$\frac{(1+Q)}{(1-Q)} = \frac{\left[1 + (\beta/2)\left(3\cos^2\theta_p - 1\right) + \left(\delta + \gamma\cos^2\theta_p\right)\sin\theta_p\right]}{\left[1 + (\beta/2)\left(3\cos^2\theta_p - 1\right) - \left(\delta + \gamma\cos^2\theta_p\right)\sin\theta_p\right]}.$$

An analysis of the effect of these non-dipolar effects on the photoemission quantity measured in NIXSW that includes the effect on the coherent interference of the incident and reflected X-rays has been presented by Vartanyants and Zegenhagen (1999) for the case of emission from initial *s*-states. In this case, not only does the electric dipole interaction lead to only a single outgoing partial (*p*-) wave ($l = 1$), but the electric quadrupole interaction also leads to a single outgoing *d*-wave ($l = 2$) while the magnetic dipole term is zero ($\delta = 0$). In the nomenclature introduced above this can be written as

$$\frac{d\sigma}{d\Omega} \propto \left[1 + R\frac{(1+Q)}{(1-Q)} + 2\sqrt{R}f_{co}\frac{(1+Q^2\tan^2\Delta)}{(1-Q)}\cos\left(\Phi + \psi - \frac{2\pi D}{d_H}\right)\right].$$

Replacing the absorption profile of equation (4.4), where $\psi = \tan^{-1}(Q\tan\Delta)$ and $\Delta = \delta_d - \delta_p$, δ_d and δ_p being the partial phase shifts of the emitter potential for outgoing *p* and *d* waves. Δ is therefore the phase difference between the outgoing waves excited by the electric quadrupole and electric dipole terms in the matrix element. For photoemission from these initial *s*-states, $Q = \gamma\sin\theta_p/3$, while tabulated theoretically computed values of γ are readily available (Nefedov *et al* 2000). Notice, though, that Q can be determined experimentally from a relatively simple NIXSW investigation if a surface containing the absorber atom of interest can be formed with a coherent fraction of zero (Jackson *et al* 2000b). In this case,

$$\frac{d\sigma}{d\Omega} \propto \left[1 + R\frac{(1+Q)}{(1-Q)}\right].$$

Of course, as stressed above, it is not really possible to form a single atomic or molecular layer that has zero coherent fraction, but a polycrystalline multilayer can be expected to meet this condition.

The ability to extract chemical-state-specific adsorbate structures using photoemission-monitored NIXSW is illustrated by an investigation of the adsorption and dissociation of methanethiol, CH_3SH, on Cu(111). Figure 4.19 shows S 2p photoemission spectra recorded from a Cu(111) surface exposed to methanethiol at 124 K and from this surface following brief heating to successively higher temperatures. Notice that these 2p spectra comprise spin–orbit split doublets with a splitting of 1.18 eV between the $2p_{3/2}$ and $2p_{1/2}$ peaks, so each distinctly different S chemical state is represented by two peaks. Four distinct states are identified. After heating to

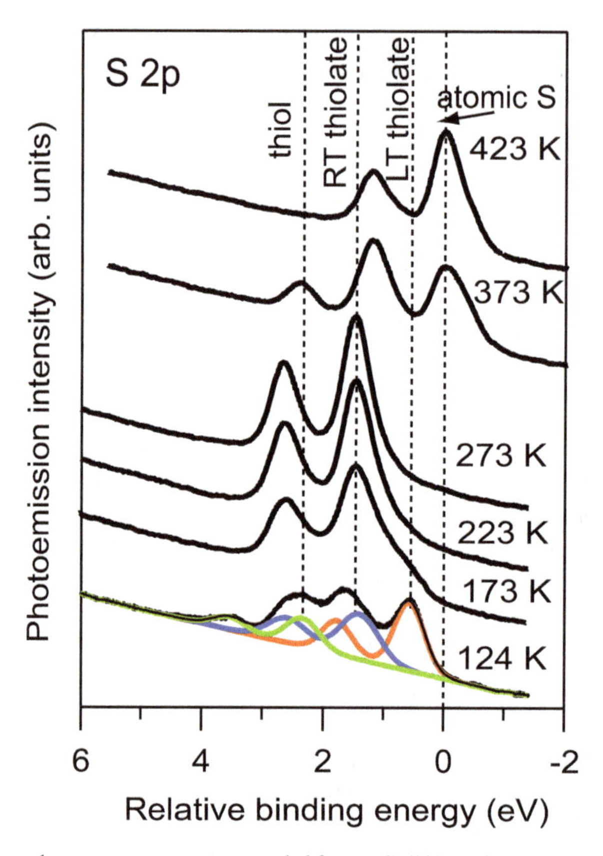

Figure 4.19. S 2p photoelectron energy spectra recorded from a Cu(111) surface onto which methanethiol was deposited at low temperature and subsequently heated briefly to increasing temperatures. Reproduced from Karriaper *et al* (1998).

the highest temperature only atomic sulphur (S) is present on the surface; the binding energies shown are relative to the value for the S $2p_{3/2}$ peak from the atomic sulphur at ~162 eV. At the lowest temperature one component with the largest binding energies is attributed to adsorbed intact thiol, while the intermediate two species are assigned to de-protonated thiolate (CH_3S-species) that differ in the way they are bound to the surface—these are referred to as LT thiolate, the dominant component at the lowest temperatures, and RT thiolate, which is the single surface species on the surface at room temperature. A fit to the lowest-temperature spectrum is shown with the thiol (green), LT-thiolate (red) and RT-thiolate (blue) components.

These same chemical shifts are found in S 1s photoemission spectra, so a set of such spectra recorded at each photon energy in a scan through the (111) normal incidence Bragg peaks allows the local adsorption geometry of each species to be determined. A subset of these spectra, recorded by Jackson *et al* (2000a), from a methanethiol-dosed Cu(111) surface at 140 K is shown in figure 4.20. Fits to each of these spectra associated with the intact thiol and the LT- and RT-thiolate species are superimposed on the experimental spectra. Clearly the overall spectral shape

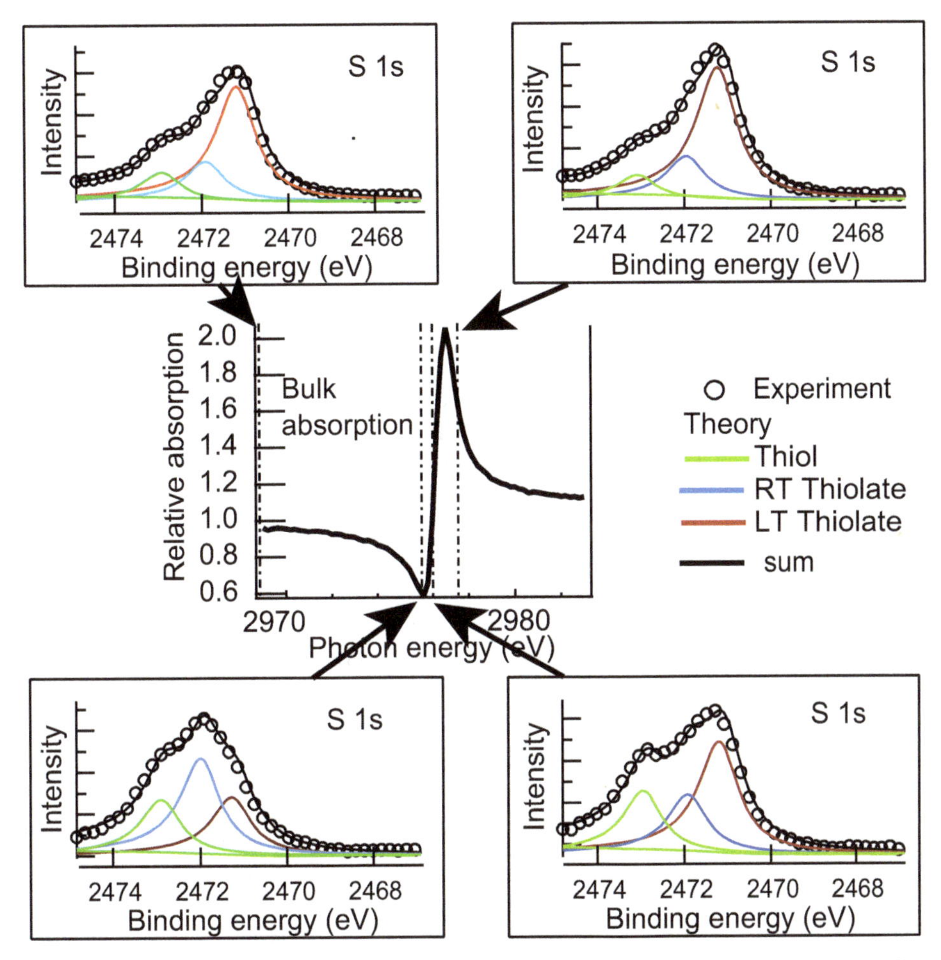

Figure 4.20. A subset of data recorded from a NIXSW study of a methanethiol-dosed Cu(111) surface at 140 K by Jackson *et al* (2000a). In the centre is shown the NIXSW absorption profile of the Cu(111) bulk. At specific photon energies, as indicated by the arrows, are S 1s photoelectron energy spectra, with fits to the different chemically shifted component species.

changes very significantly as the photon energy is changed due to the different energy dependence of the heights of the different spectral components, attributable to the different local geometry of the S atoms in these species.

(111) NIXSW photoemission profiles from these data, together with a matching set of $(\bar{1}11)$ NIXSW photoemission profiles, could therefore be extracted for each species, allowing their local geometries to be established. The results showed that the S atoms of the intact thiol occupy sites atop the surface Cu atoms, whereas for the LT thiolate the results favour hollow site occupation. The key difference between the LT- and RT-thiolate species is attributed to a complex irreversible surface reconstruction that occurs as the temperature is raised.

In discussing the technique of SXRD earlier in this chapter, attention was drawn to the *phase problem*, which is common to all XRD studies. Specifically, if one could measure both the amplitude and phase for diffracted beams it would be possible to invert these data by Fourier transforms to produce a real-space image of the structure. In practice, however, one can measure only the intensity, so the phase information is lost. However, an XSW measurement does allow one to determine the phase, albeit only for a single Fourier component of the structure corresponding to the specific diffraction condition, $H = hkl$. XSW measurements at a suitable selection of different H values can therefore be used to create the required real-space image of the structure. An early interesting example of this approach is a study by Zhang *et al* (2004) of the interface of rutile $TiO_2(110)$ with an aqueous electrolyte, specifically to identify the local sites of Sr^{2+}, Zn^{2+} and Y^{3+} ions at the interface. In order to detect the X-ray absorption at the water interface, measurements were made of the element-specific Kα fluorescence yield that is only weakly absorbed by the thin layer of water. For each measurement at a different H one obtains both the amplitude, f_H, and phase, p_h, of the element-specific normalised density profile $\rho(\mathbf{r})$, i.e.,

$$F_H = \int \rho(\mathbf{r})\exp(2\pi i\mathbf{H}.\ \mathbf{r})dr = f_H \exp(2\pi i p_H).$$

In this example measurements were made at the (110), (111), (200), (101) and (211) diffraction conditions to construct the real-space images shown in figure 4.21.

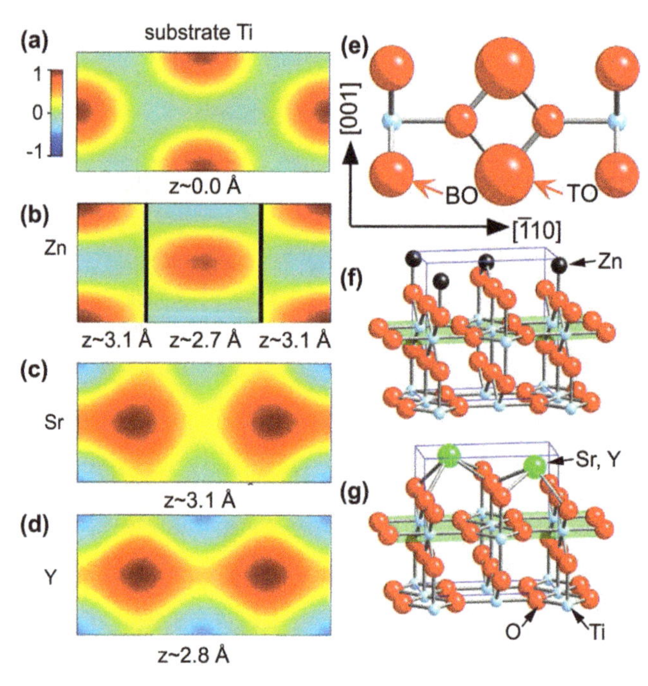

Figure 4.21. Results of XSW imaging investigation of rutile $TiO_2(110)$–aqueous electrolyte interface. Panels (a)–(d) show the lateral distributions of Ti, Zn^{2+}, Sr^{2+} and Y^{3+} in planes of maximum density at heights, z, above the TiO plane. Panel (e) is a perspective view above the rutile surface cell with the bridging oxygen (BO) and terminal oxygen (TO) atoms identified. Panels (f) and (g) show the electrolyte cation sites in perspective. Reproduced from Zhang *et al* (2004), copyright (2004), with permission from Elsevier.

Notice that the spatial resolution of these images is only ~1 Å, but they clearly identify the relevant site occupation.

4.4 X-ray absorption fine structure

As remarked in section 4.1, two surface structural techniques that rely on synchrotron radiation actually exploit the coherent interference of elastically scattered electrons rather than photons. These are SEXAFS, a surface-specific variant of the technique of EXAFS used to obtain local structural information in bulk materials, and photoelectron diffraction (PhD), which is described in section 4.5. The X-ray absorption fine structure (XAFS) technique is used extensively in synchrotron radiation experiments to obtain information on the local structural surroundings of specific atomic species in a wide range of bulk materials including compounds, alloys and biological matter. The basic measurement is of the absorption of incident X-rays as a function of photon energy above the threshold for photoionisation of the atoms of interest in the material. In its simplest form X-rays of intensity I_0 are incident on a sample of thickness t, and one measures the transmitted intensity I; the absorption coefficient is given by $\mu = -t^{-1}\ln(I/I_0)$. Figure 4.22 shows the result of such a measurement from a 1 μm Al foil as the photon energy is scanned through the Al K-edge. The sharp rise in absorption at a photon energy of ~1560 eV is due to photoionisation of the Al 1s state; the modulation ('extended fine structure') in the absorption above this threshold is the EXAFS, although it is usual to distinguish as EXAFS the 'extended' energy range of more than ~50 eV above the edge from the near-edge region (within a few tens of eV of the edge).

The photoionisation cross-section is determined by the matrix element involving the initial state $|i\rangle$ (the wavefunction of the occupied 1s state) and the final state $\langle f|$,

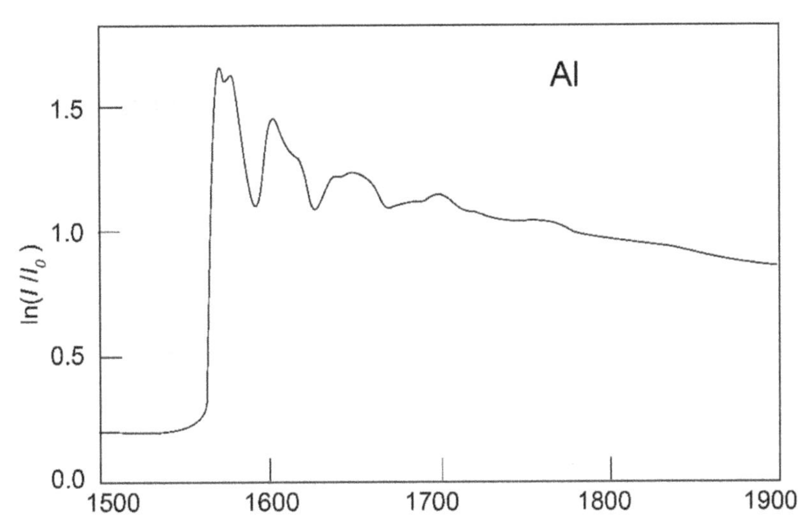

Figure 4.22. EXAFS spectrum of Al above the Al K-edge measured from a 1 μm Al foil. Woodruff (1988) John Wiley & Sons.

the outgoing photoelectron wavefield at the atom $\sigma = |\langle f|\mathbf{A}.\ \mathbf{p}|i\rangle|^2$, where $\mathbf{A}$ is the vector potential of the incident X-rays and $\mathbf{p}$ is the electron momentum operator. The EXAFS modulations arise from elastic scattering of the emitted photoelectrons that are back-scattered to the emitter, thereby determining the final-state photoelectron wavefunction at the emitter atom.

Figure 4.23 shows schematically some of the back-scattering paths that lead to changes in the photoelectron wavefunction at the emitter site through coherent interference of the outgoing wave and the back-scattered waves. Evidently, the relative phases of these components are determined by the scattering path lengths and the photoelectron wavelength (and thus the photoelectron energy). It is convenient to extract from the experimental measurement of μ, the fine structure function as a function of the photoelectron wavevector, k,

$$\chi(k) = \frac{\mu - \mu_0}{\mu_0}.$$

For the case of photoionisation of an initial s-state (as in the case of a K-edge when the ionisation is of a 1s state) from a spherically averaged (liquid, amorphous or polycrystalline) sample, and assuming that only 180° single back-scattering paths (as shown in in figure 4.23) contribute to the final state, the fine structure function can be written as

$$\chi(k) = -\sum_j \frac{N_j}{kr^2}\left|f_j(\pi, k)\right| \sin\left[2kr_j + \psi_j(k)\right] e^{-2\sigma_j^2 k^2} e^{-2r_j/\lambda_j(k)}. \tag{4.6}$$

The summation is over 'shells' of N_j atoms of the same element at a distance r_j from the X-ray-absorbing atom. $f_j(\pi,k)$ is the 180° (π) scattering factor of these atoms, $e^{-2\sigma_j^2 k^2}$ is a Debye–Waller factor dependent on the mean-square vibrational amplitudes σ_j^2, and $e^{-2r_j/\lambda_j(k)}$ is a damping term to account for inelastic scattering of the photoelectrons having a mean-free-path λ_j. The relative phase of the directly emitted and back-scattered photoelectrons is given by $[2kr_j + \psi_j(k)]$, where the first term is the phase accumulated by the back-scattering path to, and returning from, the scatterer at r_j, and $\psi_j(k)$ is the sum of the phase shifts encountered by the

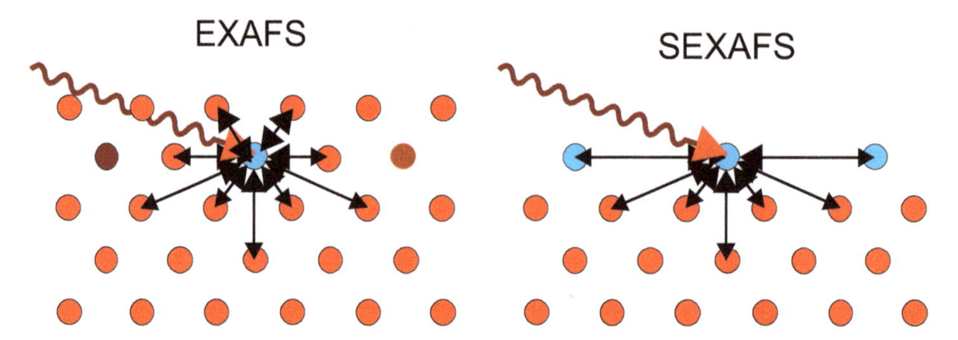

Figure 4.23. Schematic diagrams showing the photoelectron scattering paths associated with EXAFS from a bulk material and SEXAFS from a surface adsorbate.

photoelectrons in escaping from the emitter, being back-scattered by the scatterer, and re-entering the emitter. It transpires that typically $\psi_j(k) \approx ak + \delta_0$, so a Fourier transform of the fine structure function yields peak at $2r_j + a$. Evidently, a is characteristic of the elemental combination of the emitter and back-scattering atoms, so it can be determined empirically from EXAFS of a model material of the same two elements but a known structure. While this essentially empirical approach based on a Fourier transform was used in early EXAFS studies, it is now more common to base the analysis of the experimental data on model calculations for trial structures; this approach allows one to take account of multiple scattering photoelectron paths as well as the local ('spherical wave') nature of the photoelectron propagation.

As shown in figure 4.23, the underlying physics of an EXAFS study of the local structure of an adsorbate on a surface is the same as in application of the methods to a bulk material, although the number of near-neighbour scatterers will generally be less, leading to a weaker amplitude of the fine structure. The method of measurement, however, must differ: transmission through a free-standing sample is not likely to yield a detectable absorption edge from a single atomic or molecular layer at the surface. However, as discussed in the context of the NIXSW technique, the absorption, which leads to creation of core holes, can be detected by measuring one of the processes associated with the refilling of the hole holes, namely X-ray fluorescence or Auger electron emission. However, X-ray fluorescence is not surface specific, so Auger electron emission is the only direct surface-specific method of monitoring the absorption in a very thin surface layer. Notice that, unlike NIXSW, a measurement of the photoelectron yield does not offer a means of measuring variations in the total photoionisation cross-section. This is because any practical measurement of the photoelectron yield is angle resolved, and photoelectron diffraction gives rise to much stronger modulations of the partial (angle-dependent) photoemission than those of the total photoionisation cross-section of EXAFS, as discussed in section 4.4. Energy-selective detection of Auger electron emission is therefore the only *direct* method of measuring surface EXAFS, but the fact that that both photoelectrons and Auger electrons created by the photoionisation give rise to inelastically scattered and secondary electrons in the underlying solid means that monitoring the total or partial electron yield following photoionisation can offer a viable indirect route to measuring SEXAFS. Ultimately, the quality of the signal-to-noise ratio of the measured SEXAFS by one of these methods depends on the signal-to-background ratio at the absorption edge, a parameter which must often be determined empirically. Dispersive detection of Auger electron emission may lead to a larger signal-to-background ratio but to a very weak signal; total electron yield is typically much larger but the edge jump, and thus the signal-to-background ratio, may generally be inferior. While the development of applications of SEXAFS led to a significant number of studies in the 1980s and 1990s, the problem of obtaining sufficiently high-quality data has led to the technique languishing.

Nevertheless, there are some further details of the application of SEXAFS that warrant attention. In particular, while the simple formula for the fine structure function given above is based on spherical averaging of the structure as occurs in liquid, amorphous or polycrystalline samples, many surface adsorption studies are based on single-crystal substrates and the objective is often to identify the local adsorption sites on such surfaces. In this case, also for the case of an initial *s*-state, the number of atoms, Nj, in the *j*th shell is replaced by an *effective* number of atoms, N_j*, where

$$N^* = 3N(\cos^2\theta \cos^2\beta + 0.5\sin^2\theta \sin^2\beta).$$

θ is the angle between the **A** vector of the incident X-rays and the surface normal (i.e., the grazing incidence angle), while β is the angle of the vector from the absorbing atom to a scatterer atom, also relative to the surface normal (see figure 4.24).

While analysis of the SEXAFS modulations allows one to determine the distances to the neighbouring (scattering) atoms, the dependence of N^* on the angles θ and β allows one to determine the direction of the neighbouring scatterers and, thus, for an adsorbed atom, the local adsorption site. The origin of this effect lies in the angular dependence of the photoemission; for an initial *s*-state ($l = 0$) the outgoing photoelectron, determined by the dipole selection rule $\Delta l = \pm 1$, is a *p*-wave ($l = 1$) directed along the radiation **A** vector, so this acts as a (very broad) 'searchlight' illuminating scatterers preferentially in this direction.

Figure 4.25(a) shows Cl K-edge SEXAFS data from a simple investigation of this type, namely the determination of the local adsorption geometry of Cl atoms on a Cu(111) surface in an ordered $(\sqrt{3} \times \sqrt{3})$R30° overlayer. Notice that the plots are of $\chi(k)k$ rather than simply $\chi(k)$; the amplitude of the modulations in $\chi(k)$ decay with increasing k due both to the inverse dependence on k in the fine structure function and also due to the Debye–Waller factor. It is therefore common to plot $\chi(k)$ multiplied by k or by k^2 to show modulations of approximately constant amplitude over the whole investigated energy range.

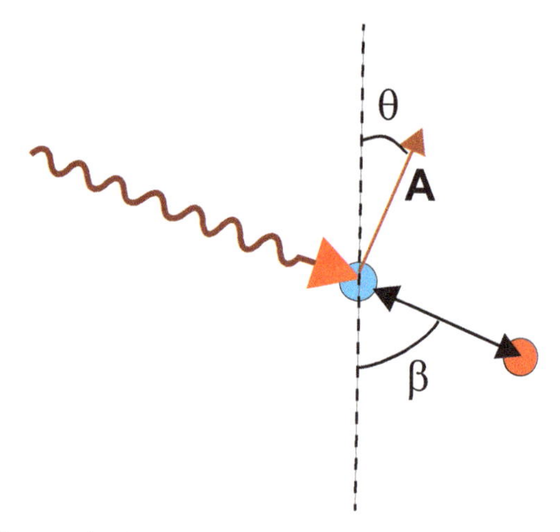

Figure 4.24. Schematic diagram showing the definition of the angles θ and β for absorption at an adsorbate atom (blue) relative to a scatterer atom (red).

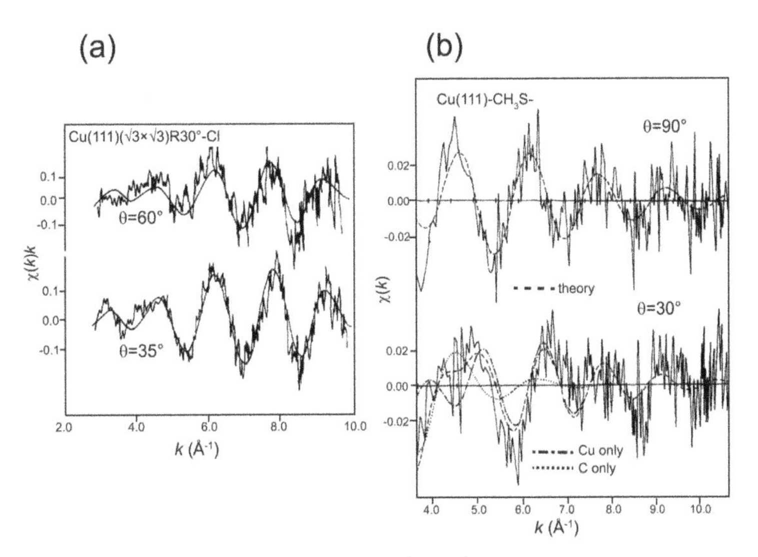

Figure 4.25. (a) Cl K-edge SEXAFS from a Cu(111) ($\sqrt{3} \times \sqrt{3}$)R30°-Cl surface at two different incidence angles. Superimposed on the raw data are back-transforms of the windowed nearest-neighbour contributions for their Fourier transforms. (b) S K-edge SEXAFS from a Cu(111)-CH_3S- surface. Superimposed on the raw data are the results of a curved wave calculation for the model structure. For the grazing incidence data the separated Cu and C back-scattering contributions are shown. Panel (a) reprinted from Crapper *et al* (1987), copyright (1987), with permission from Elsevier. Panel (b) reprinted from Prince *et al* (1989), copyright (1989), with permission from Elsevier.

Superimposed on the raw data are smooth lines that represent the dominant nearest-neighbour back-scattering contribution. These are obtained by taking a Fourier transform of the raw data, setting a window around the dominant peak in the transform, and back-transforming the windowed function. The good fit of these back-transforms to the raw data shows that the SEXAFS is dominated by these nearest-neighbour contributions, while their amplitudes as a function of θ allowed the adsorption site to be identified as a threefold coordinated hollow. In this study the Cu-Cl nearest-neighbour distance was found to be 2.39 ± 0.02 Å, the $\psi_j(k)$ phase-shift correction term being determined by comparison with EXAFS recorded from a Cu-Cl model compound. Later studies of similar systems generally achieved the data analysis with the aid of model calculations for different trial structures, taking account of multiple scattering and the 'curved wave' nature of the local photo-electron propagation—the approach now used routinely in most bulk EXAFS studies. An example of this, for S K-edge SEXAFS from Cu(111)-CH_3S-, is shown in figure 4.25(b). For the spectrum recorded at grazing incidence the separate contributions from Cu and C back-scatterers are shown; the fact that the C back-scattering is only found at grazing incidence indicates that the S-C bond is perpendicular to the surface. In EXAFS investigations of bulk materials it is generally possible to identify the contributions of several more distant scattering neighbours, but typically in SEXAFS experiments the signal-to-noise ratio and the intrinsically weak modulations due to reduced numbers of neighbours at the surface means that often only the nearest neighbours can be reliably identified.

While this simple scattering picture is used in the analysis of EXAFS, the modulations in the absorption spectrum in the near-edge region are commonly described in a different way in terms of unoccupied electronic states. Of course, it is possible to calculate the properties of these unoccupied states by scattering calculations, so the underlying physics is unchanged, but a different description is often more informative. Studies of the modulations in the near-edge range (up to ~50 eV above the edge) are commonly referred to as X-ray absorption near-edge structure (XANES) or NEXAFS in different communities and application fields, but are not consistently used in publications. NEXAFS was generally introduced to describe effects seen in molecular systems to describe electronic transitions to unoccupied quasi-bound molecular resonances, and is exploited in structural studies, particularly to determine molecular orientations, whereas XANES is the more common acronym used in the context of transitions to unoccupied electronic states in inorganic materials, as already introduced in chapter 3.

In NEXAFS, as in EXAFS and photoemission, dipole selection rules apply, so transitions from initial *s*-states must be to final *p*-states. In diatomic and simple planar molecules the molecular states can be classified as having σ or π symmetry. In the case of a diatomic molecule σ states are totally symmetric about the molecular axis, whereas π states are antisymmetric with respect to this axis. For planar molecules σ and π states are, respectively, symmetric or antisymmetric relative to the molecular plane. If NEXAFS is measured above the K-edge of a constituent atom in the molecule (i.e., from the totally symmetric 1s state) then the intensity of transitions to final states of σ- or π-symmetry are determined by the orientation of the incident **A** vector relative to the molecular axis or plane, providing a method of determining the molecular orientation.

The diatomic molecule CO, much studied as an adsorbate on metal surfaces, provides a simple example of this application of NEXAFS. The lowest unoccupied molecular orbital of CO is the 2π state, so the transition from the 1σ (O 1s) state to this final state will be excited if the incident **A**-vector has a component perpendicular to the C–O molecular axis but will not be excited if the **A**-vector is parallel to the molecular axis. Specifically, the probability of exciting the transition to this π-symmetry final state is proportional to $\sin^2\alpha$, where α is the angle between the molecular axis and the **A**-vector. Figure 4.26 shows O K-edge NEXAFS spectra recorded from CO adsorbed on Pt(533) at two different incidence angles, namely a grazing incidence with the **A**-vector at 20° from the surface normal ($\theta_A = 20°$) and a near-normal incidence ($\theta_A = 80°$); these spectra were recorded by setting the electron spectrometer to the pass energy corresponding to the O KVV Auger transition at 513 eV and scanning the photon energy.

A sharp peak just above the threshold for photoionisation corresponds to transitions to the unoccupied 2π state of the CO molecule. A second much broader peak some 15 eV higher in energy, labelled σ, corresponds to a scattering resonance to a quasi-bound state of σ symmetry. Note that this second peak is almost absent at near-normal incidence, indicating that the C–O molecular axis must be

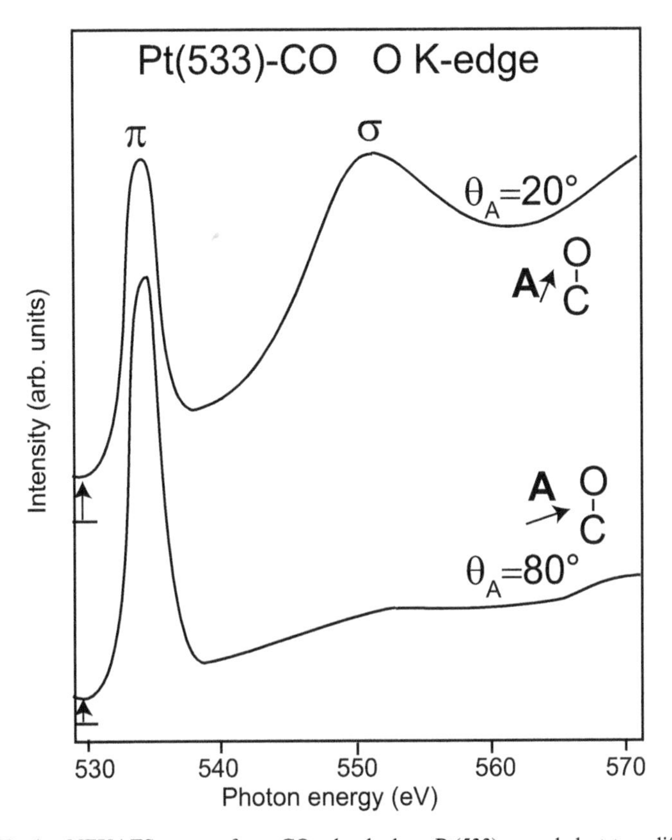

Figure 4.26. O K-edge NEXAFS spectra from CO adsorbed on Pt(533) recorded at two different incidence angles. Reprinted from Somers *et al* (1987), copyright (1987), with permission from Elsevier.

approximately parallel to the surface normal. The inset schematic sketches show the implied relationship of the **A**-vector relative to the molecular axis. Consistent with the near-perpendicular orientation of the molecule the intensity of the π-resonance (relative to the background edge jump, shown by the arrows) is maximum near normal incidence.

Pt(533) is a stepped surface with an average surface orientation of 15° relative to the (111) terraces separated by monoatomic steps. Temperature-programmed desorption spectra show that CO is adsorbed preferentially at the step sites, relative to adsorption on the (111) terraces; CO desorbs from the steps at ~100 °C higher temperature than from the terraces, so by annealing at the appropriate temperature it is possible to prepare a surface on which CO is adsorbed only at the step sites. Figure 4.27 shows the intensity of the π-resonance measured from such a surface as a function of the incidence direction. The continuous curve fitted to these data corresponds to a tilt angle of the C–O axis relative to the (533) surface normal of 5° with an estimated precision of ±7°. The significant conclusion of this study is therefore that edge site adsorption does not lead to a strongly tilted CO molecule, as had been reported for some other surfaces. Notice that this analysis is sensitive to the

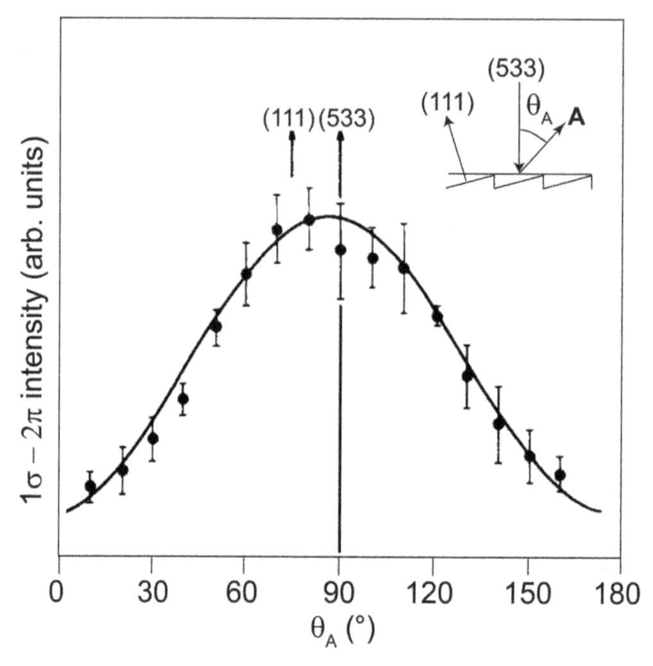

Figure 4.27. Incident polarisation angle dependence of the O 1s π-resonance from CO adsorbed only at step sites on Pt(533). The fit to the data corresponds to a C–O axis tilt angle relative to the (533) surface normal of 5°. Reprinted from Somers *et al* (1987), copyright (1987), with permission from Elsevier.

degree of linear polarisation of the incident radiation, which in this case was found to be 85%.

Figure 4.28 shows an example of a NEXAFS determination of the orientation of a planar molecule, specifically the formate species, HCOO, formed on Cu(110) by interaction with (and deprotonation of) formic acid, HCOOH. The fact that the Cu (110) has only twofold rotational symmetry allows one to exploit the full impact of the NEXAFS symmetry rule by varying not only the angle of the incident **A**-vector relative to the surface normal, but also its azimuthal orientation relative to the surface. In particular, the π-resonance is most strongly excited when the **A**-vector is perpendicular to the molecular plane and is not excited when the **A**-vector lies within this plane.

The spectra of figure 4.28 thus clearly show that the molecular plane is perpendicular to the (110) surface and lies within the ($\bar{1}$10) azimuth.

4.5 Photoelectron diffraction

The EXAFS technique exploits the coherent interference of an outgoing photoelectron wavefield and elastically back-scattered components of the same wavefield at the emitter atom. As such, the emitter is both the source of the photoelectrons but also the detector, and the fact that it samples the immediate structural environment in all directions means that it is intrinsically a technique to obtain structural information from within a bulk material. By contrast, if one detects this coherent interference of the emitted and scattered components of the photoelectron wavefield

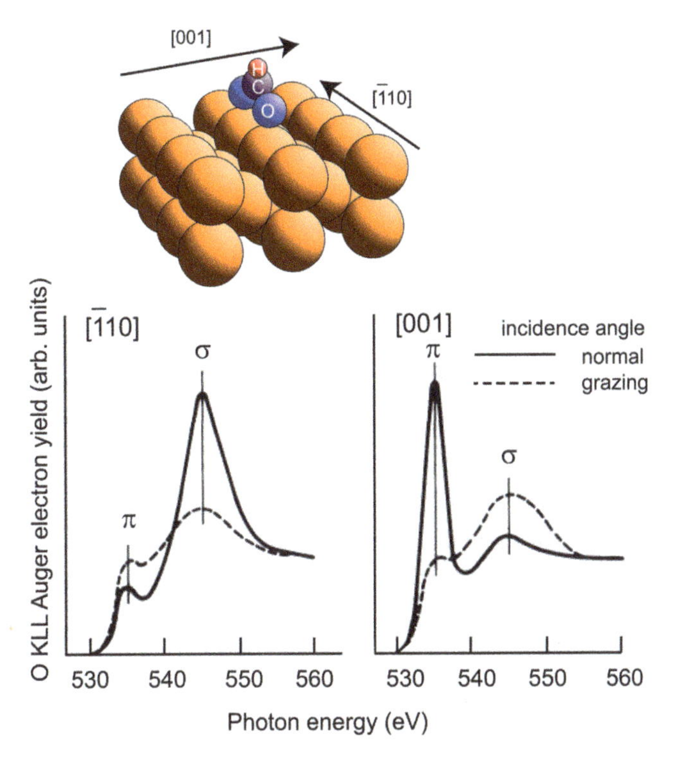

Figure 4.28. O K-edge NEXAFS recorded from the formate (HCOO) species adsorbed on Cu(110) together with a schematic representation of the structure. Reprinted figure with permission from Puschmann *et al* (1985), copyright (1985), by the American Physical Society.

outside the solid, the resulting technique is intrinsically surface specific. This technique of photoelectron diffraction has a number of important advantages over SEXAFS for surface structure determination.

Figures 4.29(a) and (b) show schematically the photoelectron scattering paths that contribute to the coherent interference in photoelectron diffraction in two different experimental geometries. Specifically, figure 4.29(a) shows the directly emitted path and the scattering paths that contribute to the detected photoemission outside the surface when core-level photoemission is excited from an atom adsorbed on the surface. The scattering path differences, which determine the relative phases and thus the interference between them, are a function of the location of the emitter atom relative to the scattering atoms and the direction of detection. The angular distribution of the detection emission is thus directly related to the local emitter site and its scattering atoms' environment. This measured intensity is also influenced by the photoelectron energy, and thereby the photoelectron wavelength, which defines the phase differences determined by the path-length differences. The structural information in photoelectron diffraction can therefore be obtained either by measuring the angular distribution of the emitted photoelectrons at a fixed photon (and photoelectron) energy, or by measuring the photoemission in a specific direction as a function of the photoelectron energy. This energy-scan mode of

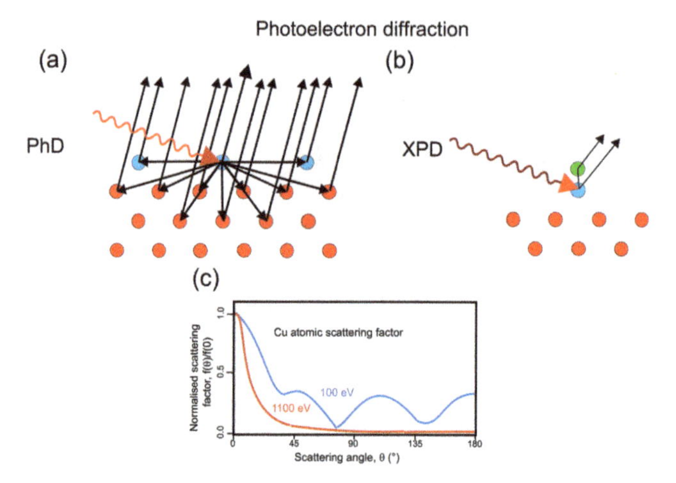

Figure 4.29. Panels (a) and (b) show the elastic scattering paths most relevant to photoelectron diffraction (PhD) in back-scattering and forward-scattering experiments. Panel (c) shows the relative elastic scattering cross-section of a Cu atom as a function of scattering angle at low and high energies.

data collection, in some ways analogous to an EXAFS measurement, is commonly given the acronym PhD. The complete angular dependence of the emission at a fixed energy is effectively a photoelectron hologram.

As is clear from figure 4.29(a), essentially all the scattering paths that are exploited to determine the location of an adsorbate relative to the underlying surface involve back-scattering, i.e., scattering angles greater than 90°, with those involving scattering angles close to 180° being particularly sensitive to near-neighbour interatomic distances. Figure 4.29(c) shows the dependence of the atomic elastic scattering cross-section of Cu as a function of scattering angle for two different electron energies. At an energy of 1100 eV the scattering cross-section is strongly peaked in the forward (0° scattering angle) direction, whereas at 100 eV this effect is much weaker; the scattering cross-section also shows a (somewhat smaller) peak at 180° back-scattering, and at some intermediate angles. This qualitative behaviour of scattering from Cu is typical, in that strong back-scattering is only found at relatively low electron energies, whereas at higher energies (>~500 eV) forward-scattering dominates. An obvious implication is that to exploit PhD to determine the location of adsorbate species relative to the underlying substrate one needs to work at relatively low photoelectron energies. This same consideration is relevant to the traditional benchmark technique of surface structure determination, namely LEED, which also exploits elastic back-scattering and is typically performed at energies of no more than ~300 eV.

Figure 4.29(b) shows a rather different surface geometry in which forward-scattering (at higher energies) can be exploited in photoelectron diffraction. The specific case illustrated is of a diatomic molecule adsorbed on the surface, high-energy photoelectron diffraction within the molecule providing information particularly on the molecular orientation. This type of high-energy forward-scattering is also potentially valuable in studies of epitaxial growth. Experiments exploiting this effect by measuring the angular dependence of the emission have generally been performed using standard laboratory XPS instruments with Al Kα and Mg Kα X-ray sources with photon energies in excess of 1 keV. This form of photoelectron diffraction is generally referred to as X-ray photoelectron diffraction (XPD). Generally, there is little if any advantage to the use of synchrotron radiation for XPD, whereas for the back-scattering low-energy experiments, and especially the scanned-energy PhD experiments, synchrotron radiation is essential, both to access the required range of photoelectron energies but also to be able to scan these energies over the range of interest.

In order to collect an experimental PhD spectrum one must measure the core-level photoemission spectrum of interest at a series of photon energies. Each spectrum must then be fitted with a peak and a background, the peak areas, $I(E)$, then being plotted as a function of photoelectron energy. Finally, following the EXAFS approach, the PhD modulation spectrum is defined as

$$\chi(E) = \frac{I(E) - I_0}{I_0}.$$

In principle I_0 is the variation in the atomic cross-section but including instrumental effects, although in practice it is simply a stiff spline through the experimental data. Figure 4.30 shows an example of a raw data set and the resulting PhD modulation spectrum for the case of N 1s emission along the surface normal from NH_3 adsorbed on Ni(111). Notice that the modulations have an amplitude of up to $\sim\pm30\%$, an order of magnitude larger than the SEXAFS modulations in figure 4.25.

To understand this it is helpful to compare the EXAFS equation (4.6) with the equation for PhD based on the same approximations, namely plane-wave single scattering and emission from an initial s-state. The PhD intensity can then be written as

$$I(k) \propto \left| \cos\theta_k + \sum_j \frac{\cos\theta_r}{r_j} f_j(\theta_j, k) W(\theta_j, k) \exp(-L_j/\lambda(k)) \exp(i(kr_j(1 - \cos\theta_j) + \delta_j(\theta_j, k)) \right|^2.$$

The summation is over scatterers at r_j.

Figure 4.31 shows the definition of the different θ angles, $\cos\theta_k$ accounting for the polarisation angular dependence of the directly emitted p-wave. $W(\theta_k, k)$ is a Debye–Waller factor, and $\exp(-L_j/\lambda(k))$ accounts for the attenuation due to inelastic scattering. The final term determines the relative phase of each scattered component due to the scattering path length, $r_j(1 - \cos\theta_j)$, and due to the passage through the emitter and scatterer potential, δ_j. Expanding this expression leads to a number of terms describing the interference of each scattered component with the directly emitted wave, together with a large number of cross-terms that may be expected to average to zero. In this case, an approximate expression for the PhD modulation function is

$$\chi_{\mathrm{PhD}}(k) = \sum_j \frac{\cos\theta_r}{r_j} f_j(\theta_j, k) W(\theta_j, k) \exp(-L_j/\lambda(k)) \exp(i(kr_j(1 - \cos\theta_j) + \delta_j(\theta_j, k)). \tag{4.7}$$

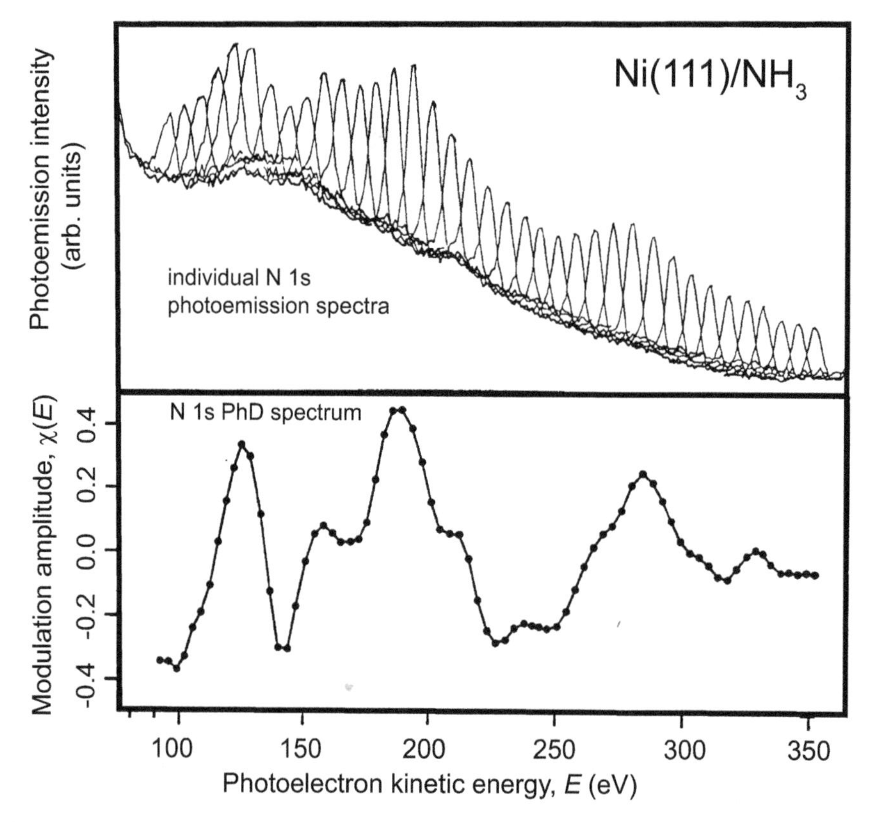

Figure 4.30. Data processing of individual normal emission N 1s spectra recorded at different photon energies from a Ni(111)-NH_3 surface to produce a N 1s photoelectron diffraction (PhD) modulation spectrum. Reprinted figure with permission from Schindler *et al* (1992), copyright (1992), by the American Physical Society.

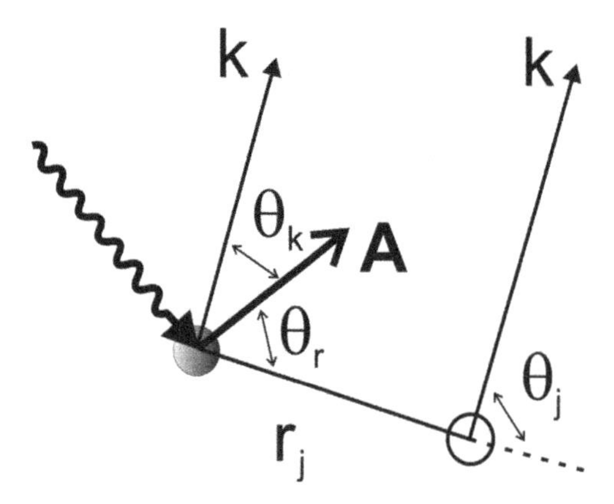

Figure 4.31. Local geometry of the emitter and one scatterer in photoelectron diffraction showing the definition of the θ angles.

To be compared with the equivalent EXAFS expression:

$$\chi_{\text{EXAFS}}(k) = -\sum_j \frac{N_j}{kr^2}\left|f_j(\pi, k)\right| \sin\left[2kr_j + \psi_j(k)\right] e^{-2\sigma_j^2 k^2} e^{-2r_j/\lambda_j(k)}. \tag{4.8}$$

A key difference is that whereas the PhD amplitude has a denominator r_j, the denominator in the EXAFS expression is kr_j^2, while r_j will take values >~2.5 Å and k is in the range ~4–10 Å^{-1}. The resulting reduction in amplitude is partially offset by the N_j in the numerator of the EXAFS expression, but still accounts for the EXAFS modulation amplitudes being smaller by a factor ~10 than those in PhD. In effect, EXAFS is PhD averaged over 4π steradians of angle. The intrinsically stronger modulations of PhD are clearly also an advantage of this technique over SEXAFS for which, as discussed above, achieving an adequate signal-to-background signal can be challenging.

Notice that while in XSW it is possible to use photoemission as a monitor of relative photoabsorption, a photoemission measurement could only be used to monitor EXAFS ('PEXAFS') if it were possible to average the photoemission over 4π steradians. Indeed, the fact that a PhD measurement is not angularly averaged but corresponds to a real-space directional measurement is a major advantage in determining a surface structure, because the emission direction used to record a PhD spectrum probes the structure preferentially in this direction. In particular, if the emission direction corresponds to 180° back-scattering from a substrate nearest neighbour to the emitter, the PhD modulation spectrum is typically dominated by the interference associated with this one back-scattering path. This arises in part due to the peak in the scattering cross-section at 180° and in part due to the local symmetry of the scattering environment. This effect is evident in the N 1s PhD spectrum of figure 4.30; these data correspond to normal emission and the nitrogen (N) atom is found to occupy an atop adsorption site, so the dominant modulation that is approximately periodic (in k, i.e., in $\sqrt{E}$) can be largely attributed to this 180° nearest-neighbour back-scattering path.

In view of this it is tempting to believe that Fourier transforms of PhD spectra can provide a direct route to interpreting the data in terms of the real-space structure, as can be fruitful in EXAFS. Indeed, early scanned-energy-mode PhD modulation spectra from the Shirley group, who pioneered this mode of operation, were referred to as angle-resolved photoemission extended fine structure (ARPEFS); see Barton *et al* (1986). Comparison of equations (4.6) and (4.7) shows one of the complications of this approach. A Fourier transform of the EXAFS expression yields the distance from the emitter to the jth scatterer (and back) $2r_j$ plus a phase shift associated with the 180° scattering process. However, a similar transform of the PhD equation yields scattering path lengths $r_j(1 - \cos\theta_j)$ with different values of θ_j and $f(\theta_j, k)$ for each scatterer. Of course, if the one nearest-neighbour 180° scatterer is sufficiently dominant, one obtains a result much more similar to that of EXAFS. Collecting PhD spectra over a wide range of emission directions can allow one to identify geometries that correspond most closely to a nearest-neighbour 180° scattering condition, leading to a modulation spectrum dominated by a single periodicity in k. Fourier transforms can help to identify these directions and form the basis of the 'projection method' (Hofmann *et al* 1994), which has been applied to more than 30 experimental data sets (Woodruff *et al* 2001), identifying both its strengths and limitations.

As remarked above, the angular distribution of photoemission resulting from photoelectron diffraction can be regarded as being a photoelectron hologram, as first pointed out by Szoke (1986). In order to exploit this observation, of course, one needs to have a method of inverting the hologram into a real-space image. Unlike an optical hologram, formed by visible radiation, there is no experimental route to this reconstruction using radiation of a similar wavelength to that used in constructing the hologram, so one must rely on numerical methods. Barton (1988) was the first to pick up on Szoke's observation and explore a Fourier transform type of approach, while there followed many attempts to produce alternative variations on this idea. However, there have been very few attempts to apply these methods to back-scattering experimental data, most reports relying on simulated data. Some more recent successes have actually come from forward-scattered photoelectron diffraction data at intermediate energies of a few hundred eV; most of these have been of the essentially bulk structure of the near-surface region (Matsushita *et al* 2010) but have included some near-surface structures (Tsutsui *et al* 2017). A realistic goal of these direct methods is not to provide a highly precise structural solution, but to provide images that identify the most likely local geometries and adsorption sites that can then be refined by more exact data-simulation methods.

In particular, any precise structure determination from PhD (and EXAFS) data must take account of the role of multiple scattering, which is not included in the simplified approach as used to derive equations (4.6) and (4.7). Notice that the electron scattering energy range of both EXAFS and PhD, from a few tens of eV to a few hundred eV, is essentially the same as that of the traditional surface structural benchmark technique of LEED. In LEED, in which the electron source is outside the surface and can be represented by a plane wave, one of the main reasons for its surface specificity is the large elastic atomic scattering cross-sections in this energy range. This helps to ensure that the diffracted beam intensities are dominated by scattering from the outermost few layers, but also means that multiple scattering contributes significantly to these intensities. Because of this, a crucial ingredient of a LEED surface structure determination is multiple scattering simulations of the diffracted intensities, the structural solution being identified by comparison of the experimental data with these calculated intensities for a succession of trial structures.

These same considerations, and ultimately the same analysis strategy, apply to PhD and EXAFS, although the role of multiple scattering is reduced relative to that of LEED. This can be attributed to the fact that these techniques use a local source of electrons, namely the atom the geometry of which is to be determined. While low electron energies ensure that scattering cross-sections are large, it is also true that back-scattering is weaker than forward-scattering (as seen in the example of figure 4.29(c)) so the dominant multiple scattering paths are those that involve only one back-scattering event in addition to forward-scattering. In EXAFS, in which all the relevant scattering events must return to the emitter, the most important multiple scattering paths are those that arise when a near-neighbour scatterer is in line with a more distant scatterer such that the more distant scatterer is 'illuminated' by 0° forward-scattering in addition to the directly emitted wave component, and indeed the same forward-scattering occurs on the return path. Taking account of these processes is essential in precise analysis of EXAFS data that shows a significant contribution from more distant neighbours. The multiple scattering simulation software packages for detailed EXAFS data analysis also take account of the local 'curved wave' character of the electron; equations (4.6) and (4.7) are based on the simpler plane-wave approximation. In practice, for the special case of SEXAFS, in which the weaker contributions from more distant scatterers are often not detectable, multiple scattering is generally less important.

Multiple scattering is much more important in PhD. This technique is, in effect, a local form of LEED, though the local electron source means that it is typically sufficient to include multiple scattering to a lower order than is required in LEED. Notice that using the external plane-wave electron source, the dominant LEED path-length differences that are sensitive to the location of an atom involve differences in scattering paths between paths from the atom of interest and paths from its neighbours. By contrast, in PhD the path-length differences are between the (emitter) atom of interest and its neighbours. This makes PhD intrinsically more sensitive to the local atomic site than LEED. Of course, an even more important advantage of PhD is that the data are specific to a particular elemental (and chemical-state) species, whereas LEED is only sensitive to different elemental character through differences in the scattering cross-sections. Like XSW, and unlike SXRD and LEED, PhD is sensitive to the *local* geometry of surface atoms and is not dependent on them having long-range order. Ultimately, a full structure determination by PhD relies on a similar trial-and-error search of model structures for which multiple scattering simulations give the best agreement with experiment. As in LEED, a reliable analysis requires the use of an objective measure of the quality of agreement between theory and experiment, and a sufficiently large data set of measurements in a range for different directions (different diffracted beams in LEED, different emission directions in PhD).

The quality of the agreement between experimental (χ_{exp}) and theoretical (χ_{th}) modulation functions in a PhD study is generally judged by the value of a reliability or R-factor, defined as

$$R = \frac{\sum(\chi_{th} - \chi_{exp})^2}{\sum(\chi_{th}^2 + \chi_{exp}^2)},$$

where the summations are over all energies and emission directions of the data set. Perfect agreement corresponds to $R = 0$, a value of 1 indicates uncorrelated data and a value of 2 corresponds to anticorrelated data. Alternative structural models are tested to identify the one that gives the lowest value of R, experience with this approach providing a guide to what is an acceptably low value to indicate a correct solution. The amount by which R increases with small changes in individual structural parameters provides an estimate of the precision with which each parameter is determined. The need for a sufficiently large data set for a correct solution is illustrated by an early investigation of the structure of PF_3 adsorbed on Ni(111) by Dippel *et al* (1983). Initial comparison of the experiential PhD modulation spectrum measured at normal emission to the surface with simulated spectra for different local adsorption sites (atop, bridge, hollows) and different heights above the surface clearly identified the atop site (figure 4.32(b)) as favoured but, as seen in figure 4.32(a), the R-factor shows minima at several different P–Ni spacings. Figure 4.32(b) shows the comparison of the experimental and theoretical spectra for these three alternative structures. As the P–Ni distance is increased, the periodicity of the predicted modulations decreases, such that there is an approximate match of the main periodicity at the three different P–Ni distances, although the best

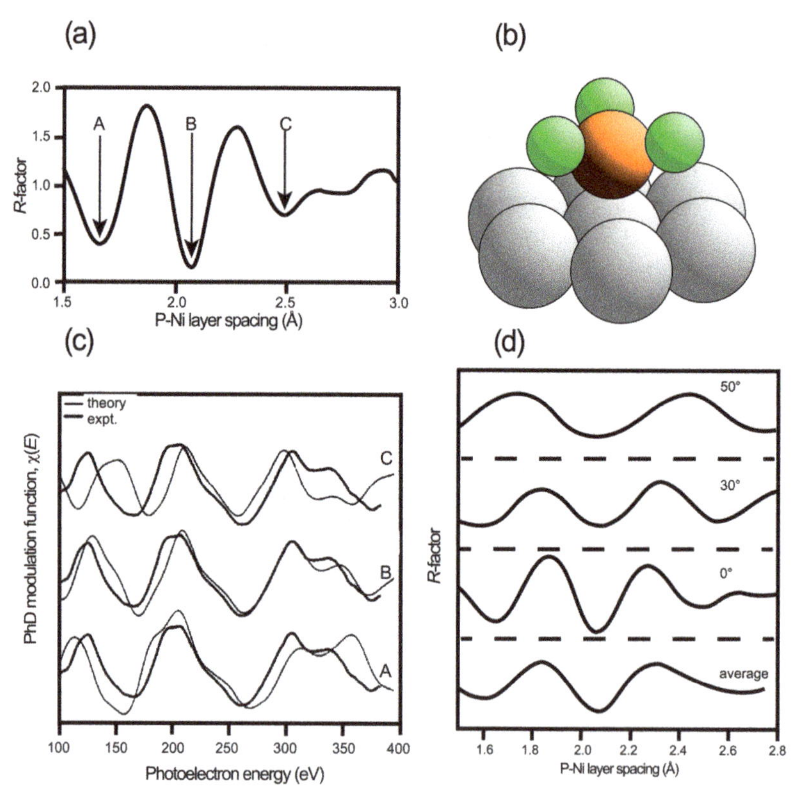

Figure 4.32. P 1s PhD results from PF_3 adsorbed on Ni(111). Panel (a) shows the variation of R-factor resulting from a comparison of the normal emission PhD spectrum with simulated spectra for atop adsorption (b) at different P–Ni distances. Panel (c) shows the comparison of the experimental and simulated spectra at the spacings A, B and C in (a). Panel (d) shows the variation of R-factor with P–Ni distance for off-normal emission spectra. Panels (a), (c) and (d) reprinted from Dippel *et al* (1983), copyright (1983), with permission from Elsevier.

fit evidently corresponds to the minimum marked as B. Enlarging the data set to include spectra recorded at two different polar emission angles clearly identifies this as the correct structure. At the off-normal emission angles the change in P–Ni distance also leads to multiple minima in the *R*-factor (figure 4.32(d)), but the different path lengths in these geometries lead to minima at different P–Ni distances. Only the value at minimum B (2.07 ± 0.03 Å) is present for all three emission directions, clearly identifying this as the optimum solution.

While this simple example illustrates the general methodology of the PhD technique, it fails to exploit the chemical-state specificity that allows it to be used for significantly more complex adsorbate systems. One example of this is provided by investigations of the adsorption of the nucleobases cytosine, thymine and uracil on Cu(110) (Jackson *et al* 2010, Allegretti *et al* 2007, Duncan *et al* 2011); these molecular structures are shown in figure 4.33. The key questions are how these molecules interact with the copper surface, what is their orientation and what is the local adsorption site.

Initial information comes from the core-level photoemission from the constituent atoms. Figure 4.34 shows O 1s and N 1s spectra from uracil and thymine after adsorbing onto Cu(110) at room temperature and after annealing to ~500 °C. Prior to this annealing the N 1s spectra show two distinct peaks with a binding energy difference of ~1.7 eV, while there is only a single O 1s peak. The two N 1s peaks clearly indicate that the two N atoms in the molecule have significantly different local bonding configuration, whereas the single O 1s peak indicates that the two O atoms in the molecule have very similar bonding configurations. The two distinct N species can most obviously be accounted for if one of the N in the molecule is de-protonated; the N(3) atom could be de-protonated if the molecule then bonds to the Cu surface through this N atom and the two O atoms. Annealing leads to a strong attenuation of the lower-binding-energy N 1s peak; N 1s spectra from adsorbed and annealed cytosine show similar behaviour. This change can be attributed to

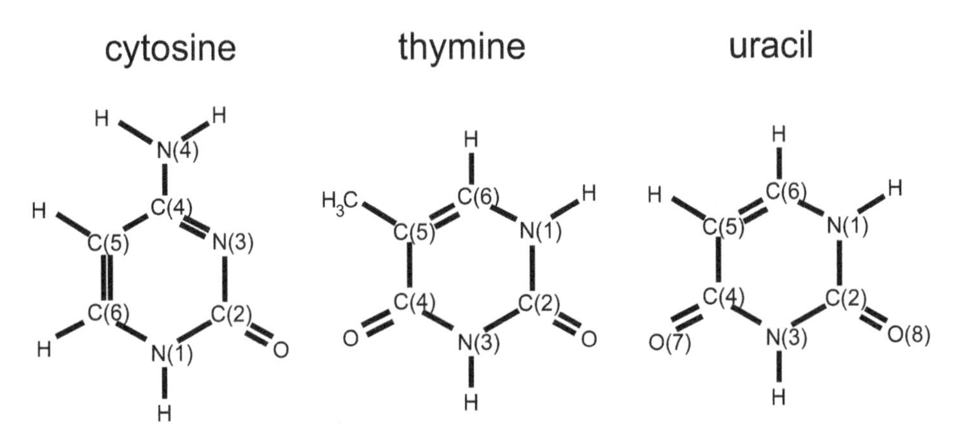

Figure 4.33. Molecular structures of cytosine, thymine and uracil showing the labelling convention used in the text for the constituent atoms.

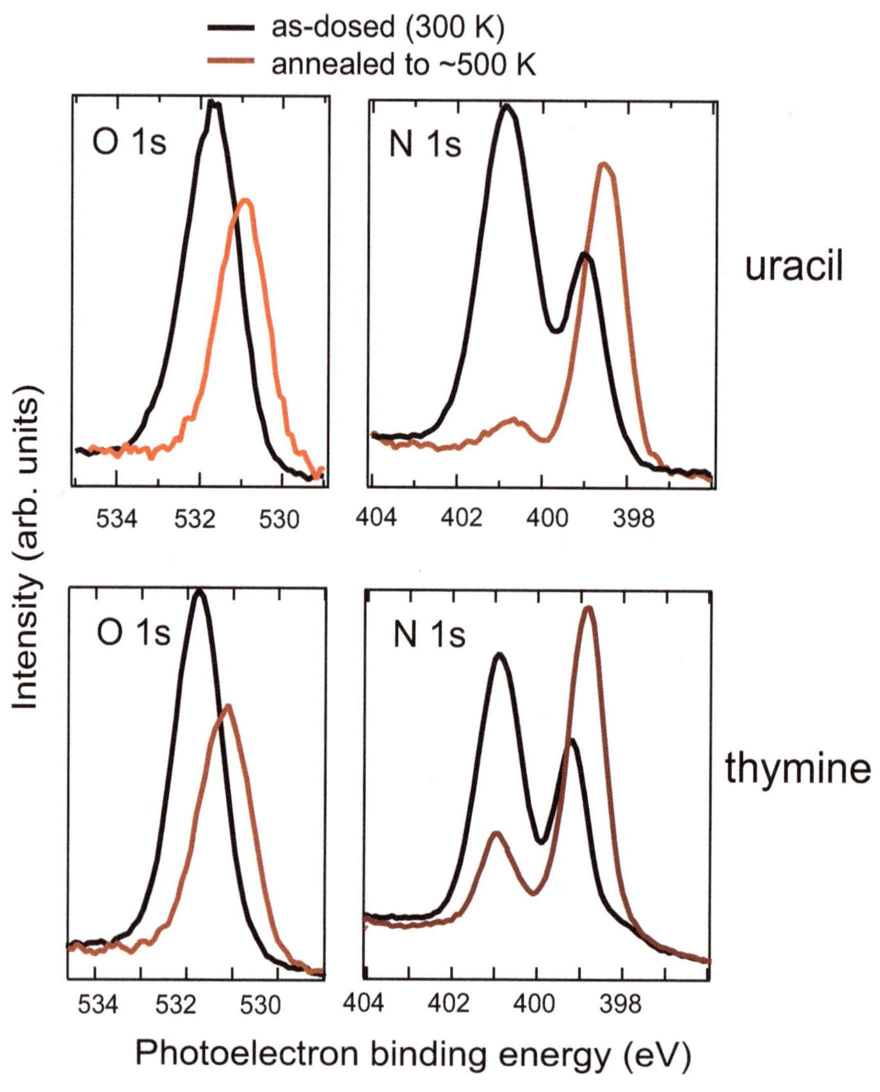

Figure 4.34. O 1s and N 1s XP spectra recorded at photon energies of 500 and 700 eV, respectively, from uracil and thymine adsorbed on Cu(110) at room temperature and after annealing to ~500 °C, reported in the studies by Duncan *et al* (2011) and Allegretti *et al* (2007).

deprotonation of the second N atom, implying that the higher-binding N 1s energy is associated with hydrogenated N. Annealing also leads to a small shift in the binding energy of the single O 1s peak for uracil and thymine. This shift of the single O 1s peak indicates that both O atoms undergo a similar change in bonding environment. The PhD data, which allow the local site of the two inequivalent N atoms to be determined, clearly show that the N(3) atom is de-protonated on initial adsorption while the N(1) atom is de-protonated on annealing.

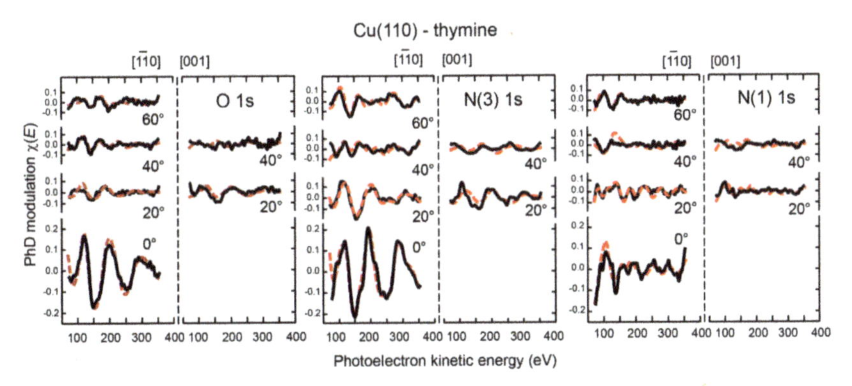

Figure 4.35. Comparison of experimental (black) N 1s and O 1s PhD spectra recorded in six different emission directions for Cu(110) after exposure to thymine and heating to ~500 °C with simulated spectra (red dashed) for the best-fit model structure in the investigation of Allegretti *et al* (2007).

The evidence for this conclusion is provided by a comparison of the experimental PhD modulation spectra from the O 1s and the two distinct N 1s photoemission peaks with simulations for different structural models; figure 4.35 shows a comparison of the experimental PhD spectra with simulated spectra for the best-fit structure. PhD spectra are shown for six different emission directions for each species, specifically four different polar emission angles in the $[1\bar{1}0]$ azimuth and two polar emission angles in the [001] azimuth. A striking feature of these data is that the normal emission spectra of the O 1s and one of the N 1s species (clearly identified as from the N(3) atom by its structural location) are not only almost identical, but also show the strongest modulations dominated by a single period in k. This behaviour is characteristic of 180° back-scattering from a near-neighbour Cu atom, clearly identifying the location of the O atoms and the N(3) atom as atop (or very close to atop) surface Cu atoms. By contrast, the PhD modulations shown by the other N 1s emitter are significantly weaker and show shorter-period modulations, indicating the influence of a longer back-scattering path length. This behaviour is consistent with this N atom being the N(1) atom that is higher above the Cu surface. Figure 4.35 shows ball models of the best-fit structure for adsorbed thymine, but also adsorbed uracil and cytosine derived from similar PhD studies. Notice that these models do not show the location of the H atoms in these molecules, as H is such a weak scatterer of electrons that the PhD technique is unable to determine their location.

These examples not only illustrate the value of the chemical-state specificity of the PhD technique to determine the location of atoms of the same element but a different bonding environment within adsorbed molecules, but further show how this structural information can help to identify which chemically shifted photoelectron binding energies correspond to which chemical species.

A rather different situation in which this chemical-state specificity can prove to be invaluable in surface structural studies is in investigations of oxygen-containing molecules on oxide surfaces. A particular application of this is in the study of the

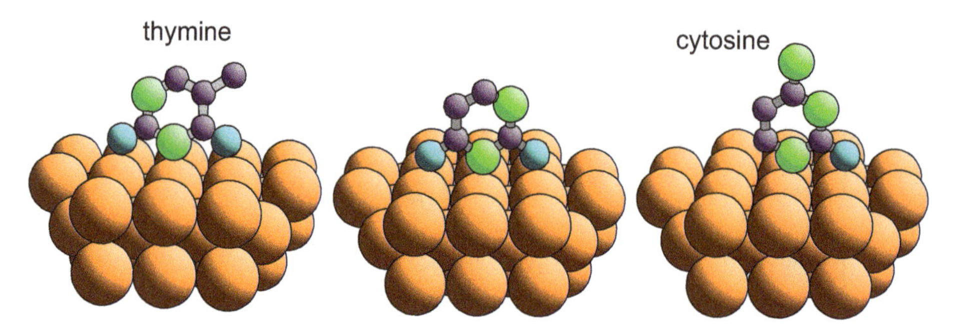

Figure 4.36. Schematic ball models of the optimum adsorption geometry of thymine, uracil and cytosine adsorbed on Cu(110) as determined by the PhD study. Atom colouring is blue (O), green (N) and black (C). H atoms are omitted from these diagrams as they are such weak electron scatterers that the PhD technique is insensitive to their presence.

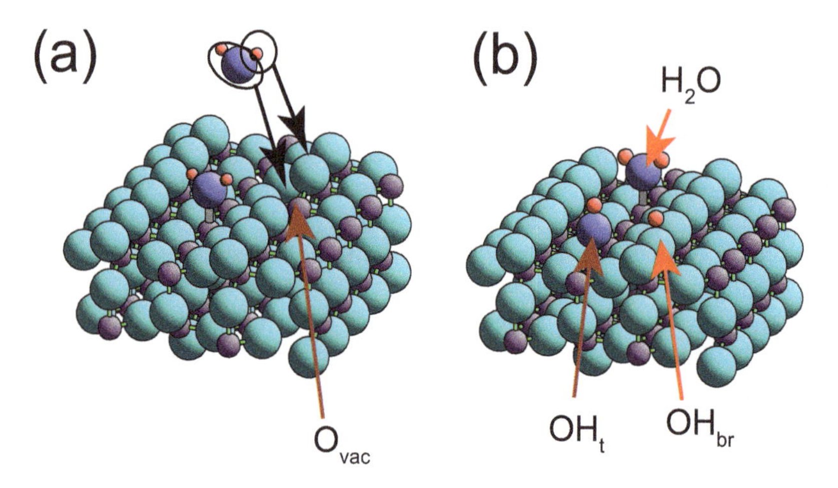

Figure 4.37. Ball model of the rutile $TiO_2(110)$ surface. Ti atoms are shown as black, O atoms as blue, but using a different shade of blue to represent the O atoms in and derived from H_2O.

interaction of water with titania surfaces, a system that has, as remarked earlier, created considerable interest since the initial discovery of electrochemical photolysis of water at this surface more than 50 years ago (Fujishima and Honda 1972). Most of this work has been on the (110) face of the rutile phase of TiO_2, shown schematically in figure 4.37. The surface is terminated by O atoms in bridging sites relative to the titanium (Ti) atoms below, but even in the best-prepared surfaces there are some of oxygen vacancy sites (O_{vac}). Many investigations have shown that exposure to water at room temperature leads to dissociation only at these O_{vac} sites; the water dissociates into an OH species that occupies the O_{vac} site and a H species that bonds to an adjacent surface O atom to create a second-surface OH species.

Intact H_2O can also adsorb on the surface at low temperature, occupying a site atop Ti atoms that lie between the rows of bridging O atoms. This location was identified by STM but confirmed and quantified (with the Ti-O_{H2O} bond length being determined) by PhD (Allegretti *et al* 2005).

Figure 4.38 shows an O 1s XP spectrum recorded from such a surface; In addition to the large peak associated with O atoms in the oxide is a smaller peak with a chemical shift of ~3.5 eV associated with molecular H_2O, but also an intermediate state associated with surface OH. This is generally attributed to bridging hydroxyl, OH_{br}, due to the interaction of water with the O_{vac} sites; this species is stable at room temperature. However, a careful temperature-dependent XPS study by Walle *et al* (2009) revealed that following low-temperature water adsorption the amplitude of the OH component decreased at the same temperature as H_2O desorption occurs, implying that there is a weakly bonded OH species that is associated with H_2O dissociation at low temperatures on perfect areas of the surface. This second OH species has been suggested to occupy a site atop surface Ti atoms. At temperatures below that associated with adsorption of intact molecular water there is therefore proposed to be two surface OH species, OH_{br} and OH_t. A PhD study (Duncan *et al* 2012) provided direct evidence of this (and a determination of the associated bond length), as shown in the data of figure 4.39.

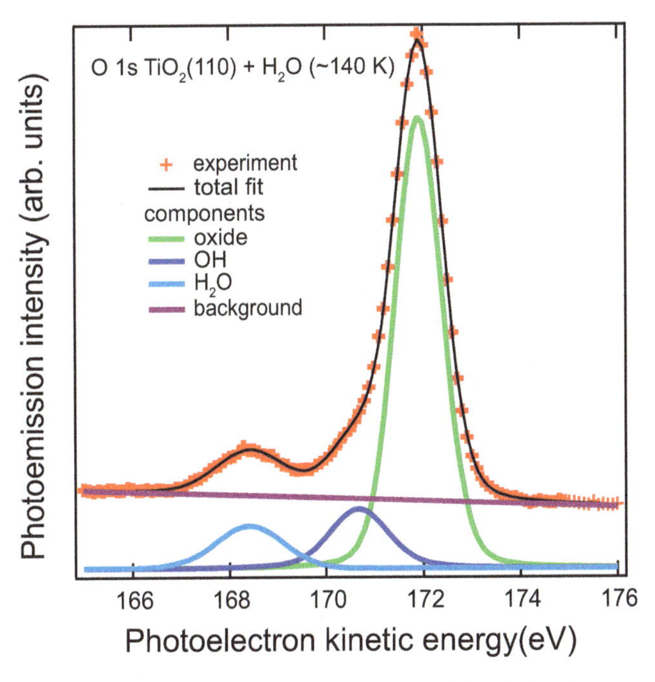

Figure 4.38. Experimental O 1s XP spectrum recorded from rutile $TiO_2(110)$ after exposure to water at a temperature of ~140 K. This is fitted by a sum of components associated with the oxide, intact water and hydroxyl OH together with a background. Data taken from the study reported by Duncan *et al* (2012).

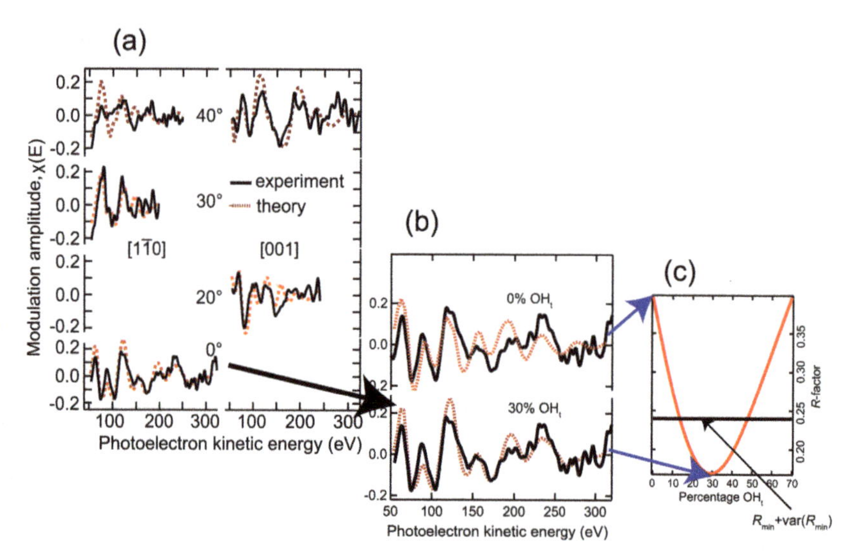

Figure 4.39. Panel (a) shows a comparison of experimental PhD spectra recorded from the OH component of the O 1s spectra in five different emission directions with the results of simulations based on coadsorption of OH_t and OH_{br} in occupations of 30% and 70%, respectively. Panel (b) shows a similar comparison of the normal emission PhD spectrum with calculations based on 0% OH_t and 30% OH_t (as for all the theory spectra in panel (a). Panel (c) shows the dependence of the R-factor on the OH_t fractional occupation in the structural model; the estimated precision of this fraction is determined by the range of values falling below the horizontal line at an R-factor value of R_{min} + var(R_{min}). Reprinted figure with permission from Duncan *et al* (2012), copyright (2012), by the American Physical Society.

References

Allegretti F, O'Brien S, Polcik M, Sayago D I and Woodruff D P 2005 Adsorption bond length for H_2O on TiO_2 (110): a key parameter for theoretical understanding *Phys. Rev. Lett.* **95** 226104

Allegretti F, Polcik M and Woodruff D P 2007 Quantitative determination of the local structure of thymine on Cu(1 1 0) using scanned-energy mode photoelectron diffraction *Surf. Sci.* **601** 3611–22

Barton J J, Bahr C C and Robey S W *et al* 1986 Adsorbate-geometry determination by measurement and analysis of angle-resolved-photoemission extended-fine-structure data: application to *c*(2 × 2)*S*/Ni(001) *Phys. Rev.* **B 34** 3807–2819

Barton J J 1988 Photoelectron holography *Phys. Rev. Lett.* **61** 1356–9

Ciston J, Subramanian A, Robinson I K and Marks L D 2009 Diffraction refinement of localized antibonding at the Si(111) 7 × 7 surface *Phys. Rev.* **B 79** 193302

Crapper M D, Riley C E and Sweeney P J J *et al* 1987 Investigation of the Cu(111) ($\sqrt{3} \times \sqrt{3}$) R30°-Cl structure using SEXAFS and photoelectron diffraction *Surf. Sci.* **182** 213

Dippel R, Weiss K-U and Schindler K-M *et al* 1983 Multiple site coincidences and their resolution in photoelectron diffraction: PF_3 adsorbed on Ni(111) *Surf. Sci.* **287/288** 465–70

Drenth J 2007 *Principles of Protein X-ray Crystallography* (New York: Springer) 3rd edn

Duncan D A, Unterberger W, Kreikemeyer-Lorenzo D and Woodruff D P 2011 Uracil on Cu (110): a quantitative structure determination by energy-scanned photoelectron diffraction *J. Chem. Phys.* **135** 014704

Duncan D A, Allegretti F and Woodruff D P 2012 Water does partially dissociate on the perfect TiO2(110) surface: a quantitative structure determination *Phys. Rev.* B **86** 045411

Faraggi M N, Jiang N and Gonzalez-Lakunza N *et al* 2012 Bonding and charge transfer in metal–organic coordination networks on Au(111) with strong acceptor molecules *J. Phys. Chem.* C **116** 24558–65

Feidenhans'l R, Grey F and Johnson R L *et al* 1990 Oxygen chemisorption on Cu(110): a structural determination by X-ray diffraction *Phys. Rev.* B **41** 5420–3

Fujishima A and Honda K 1972 Electrochemical photolysis of water at a semiconductor electrode *Nature* **238** 37–8

Gustafson J, Shipilin M and Zhang C *et al* 2014 High-energy surface X-ray diffraction for fast surface structure determination *Science* **343** 758–61

Hofmann P, Schindler K-M, Bao S, Bradshaw A M and Woodruff D P 1994 Direct identification of atomic and molecular adsorption sites using photoelectron diffraction *Nature* **368** 131–2

Hussain H, Ahmed M H M and Torrelles X *et al* 2019 Water-induced reversal of the TiO_2(011)-(2 × 1) surface reconstruction: observed with *in situ* surface X-ray diffraction *J. Phys. Chem.* C **123** 13545–50

Jackson D C, Duncan D A and Unterberger W *et al* 2010 Structure of cytosine on Cu(110): a scanned-energy mode photoelectron diffraction study *J. Phys. Chem.* C **114** 15454–63

Jackson G J, Woodruff D P and Jones R G *et al* 2000a Following local adsorption sites through a surface chemical reaction: CH_3SH on Cu(111) *Phys. Rev. Lett.* **84** 119

Jackson G J, Cowie B C C and Woodruff D P *et al* 2000b Atomic quadrupolar photoemission asymmetry parameters from a solid state measurement *Phys. Rev. Lett.* **84** 2346–9

Karriaper M S, Grom G F, Jackson G J, McConville C F and Woodruff D P 1998 Characterization of thiolate species formation on Cu(111) using soft X-ray photoelectron spectroscopy *J. Phys. Condens. Matter* **10** 8661–70

Kerkar M, Hayden A B, Woodruff D P, Kadodwala M and Jones R G 1992 An unusual adsorption site for methoxy on Al(111) surfaces *J. Phys. Condens. Matter* **4** 5043

Kostelník P, Seriani N and Kresse G *et al* 2007 The Pd(100)-($\sqrt{5} \times \sqrt{5}$)R27°-O surface oxide: a LEED, DFT and STM study *Surf. Sci.* **601** 1574–81

Kumpf C, Smilgies D and Landemark E *et al* 2001 Structure of metal-rich (001) surfaces of III–V compound semiconductors *Phys. Rev.* B **64** 075307

Lucas C A, Marković N M and Ross P N 1999 The adsorption and oxidation of carbon monoxide at the Pt(111)/electrolyte interface: atomic structure and surface relaxation *Surf. Sci.* **425** L381

Lucas C A and Marković N M 2007 *In-situ Spectroscopic Studies of Adsorption at the Electrode and Electrocatalysis* (Amsterdam: Elsevier) pp 339–81

Marković N M, Grgur B N, Lucas C A and Ross P N 1999 Electrooxidation of CO and H_2/CO mixtures on Pt(111) in acid solutions *J. Phys. Chem.* B **103** 487–95

Marks L D, Erdman N and Subramanian A 2001 Crystallographic direct methods for surfaces *J. Phys. Condens. Matter* **13** 10677

Matsushita T, Natsui F, Daimon H and Hayashi K 2010 Photoelectron holography with improved image reconstruction *J. Electron Spectrosc. Rel. Phenom.* **178–179** 195–220

Meyerheim H L, Gloege T, Sokolowski M, Umbach E and Bäurle P 2000 Adsorption-induced distortion of a large π-conjugated molecule studied by surface X-ray diffraction: end-capped quaterthiophene on Ag(111) *Europhys. Lett.* **52** 144–50

Moritz W and Van Hove M A 2022 *Surface Structure Determination by LEED and X-rays* (Cambridge: Cambridge University Press)

Mousley P J, Rochford L A and Ryan P T P *et al* 2022 Direct experimental evidence for substrate adatom incorporation into a molecular overlayer *J. Phys. Chem.* C **126** 7346–55

Nefedov V I, Yarzhemsky V G, Nefedova I S, Trzhaskovskaya M B and Band I M 2000 The influence of non-dipolar transitions on the angular photoelectron distribution *J. Electron. Spectrosc. Relat. Phenom.* **107** 123–30

Nicklin C 2014 Capturing surface processes *Science* **343** 739–40

Prince N P, Seymour D L, Woodruff D P, Jones R G and Walter W 1989 The structure of mercaptide on Cu(111): a case of molecular adsorbate-induced substrate reconstruction *Surf. Sci.* **215** 566–76

Puschmann A, Haase J, Crapper M D, Riley C E and Woodruff D P 1985 Structure determination of the formate intermediate on Cu(110) by use of X-ray-absorption fine-structure measurements *Phys. Rev. Lett.* **54** 2250–3

Robinson I K, Waskiewicz W K, Fuoss P H and Norton L J 1988 Observation of strain in the Si (111) 7 × 7 surface *Phys. Rev.* B **37** 4325–8

Saidy M, Warren O L, Thiel P A and Mitchell K A R 2001 A structural refinement with LEED for the Pd(100)–($\sqrt{5} \times \sqrt{5}$)$R27°$-O surface *Surf. Sci.* **494** L799–804

Schindler K-M, Fritzsche V and Asensio M C *et al* 1992 Structural determination of a molecular adsorbate by photoelectron diffraction: ammonia on Ni(111) *Phys. Rev.* B **46** 4836

Shipilin M, Hejral U and Lundgren E *et al* 2014 Quantitative surface structure determination using *in situ* high-energy SXRD: surface oxide formation on Pd(100) during catalytic CO oxidation *Surf. Sci.* **630** 229

Somers J S, Lindner T and Surman M *et al* 1987 NEXAFS determination of CO orientation on a stepped platinum surface *Surf. Sci.* **183** 576

Stierle A, Kasper N and Dosch H *et al* 2005 A surface X-ray study of the structure and morphology of the oxidized Pd(001) surface *J. Chem. Phys.* **122** 044706

Szoke A 1986 Short wavelength coherent radiation: generation and applications; D T Attwood and J Boker *AIP Conf. Proc. No. 147* (New York: American Institute of Physics)

Takayanagi K, Tanishiro Y, Takahashi M and Takahashi S 1985 Structural analysis of Si(111)-7 × 7 by UHV-transmission electron diffraction and microscopy *J. Vac. Sci. Technol.* A **3** 1502–6

Tsutsui K, Matsushita T and Natori K *et al* 2017 Individual atomic imaging of multiple dopant sites in as-doped si using spectro-photoelectron holography *Nano Lett.* **17** 7533–8

Van Spronsen M A, Frenken J W M and Groot I M N 2017 Surface science under reaction conditions: CO oxidation on Pt and Pd model catalysts *Chem. Soc. Rev.* **46** 4347–74

Vartanyants I A and Zegenhagen J 1999 Photoelectric scattering from an X-ray interference field *Solid State Commun.* **113** 299–320

Villegas I and Weaver M J 1994 Carbon monoxide adlayer structures on platinum (111) electrodes: a synergy between *in-situ* scanning tunneling microscopy and infrared spectroscopy *J. Chem. Phys.* **101** 1648–60

Vlieg E 2000 *ROD*: a program for surface X-ray crystallography *J. Appl. Crystallogr.* **33** 401–5

Walle L E, Borg A, Uvdal P and Sandell A 2009 Experimental evidence for mixed dissociative and molecular adsorption of water on a rutile TiO_2(110) surface without oxygen vacancies *Phys. Rev.* B **80** 235436

Wang J X, Robinson I K, Ocko B M and Adzic R R 2005 Adsorbate-geometry specific subsurface relaxation in the CO/Pt(111) system *J. Phys. Chem. Lett.* **109** 24–6
Woodruff D P 1988 From SEXAFS to SEELFS *Surf. Interface Anal.* **11** 25–35
Woodruff D P, Baumgärtel P, Hoeft J T, Kittel M and Polcik M 2001 *J. Phys.: Condens. Matter* **13** 10625
Woodruff D P and Duncan D A 2020 X-ray standing wave studies of molecular adsorption: why coherent fractions matter *New J. Phys.* **22** 113012
Zhang Z, Fenter P and Cheng L *et al* 2004 Model-independent X-ray imaging of adsorbed cations at the crystal–water interface *Surf. Sci.* **554** L95–L100

Chapter 5

Imaging

The introduction of scanning probe methods (notably scanning tunnelling microscopy and atomic force microscopy), which offer the possibility of imaging of surfaces with atomic resolution, has led to an increased awareness of the importance of inhomogeneity of even relatively idealised single-crystal surfaces. Synchrotron radiation does not enable imaging techniques with this exceptional resolution, but does offer important complementary spectromicroscopy and microspectroscopy techniques with resolutions down to a few tens of nanometres (or better) that provide information of the spatial distribution of elemental, chemical and electronic properties. These include two different methods of implementing photoelectron emission microscopy, but also spatially resolved X-ray absorption near edge structure (XANES), including methods based on circular and linear dichroism that allow imaging of magnetic and ferroelectric domains.

5.1 Introduction

In the early years of the development of 'modern surface science', following the widespread availability of commercial stainless-steel demountable ultra-high vacuum (UHV) equipment in the 1970s, which led to a wide range of new surface probes based on photons, electrons and ions, the emphasis was on studies of low-index single-crystal surfaces with an implicit assumption on their homogeneity, although there was certainly an awareness of the potential importance of different kinds of surface defects. The later development of scanning probe techniques, notably scanning tunnelling microscopy (STM) and atomic force microscopy (AFM), led to more emphasis on the role of surface heterogeneity. Of course, the main impact of these techniques was their ability to provide atomic-scale images of surfaces, although the interpretation of such images in terms of atomic coordinates and atomic identities is certainly less straightforward that it first seems. No synchrotron radiation imaging techniques offer such a high degree of spatial resolution, but at resolutions of the order of tens of nanometres several techniques provide

doi:10.1088/978-0-7503-3847-9ch5

spectroscopic information complementary to that provided by these scanning probe methods. Specifically, synchrotron radiation offers spatially dependent element-specific, chemical-state-specific, and local electronic structural information through spectromicroscopy and microspectroscopy techniques.

As discussed in chapter 2, the main objective of designs of new synchrotron radiation sources, including major upgrades of several existing sources, has been to achieve improved (reduced) emittance. The goal is often stated as being to achieve diffraction-limited emittance, which is commonly taken to mean diffraction-limited emittance in the X-ray photon energy range around 8–12 keV (1.5–1 Å wavelength). Of particular importance in achieving this goal is its impact on the use of synchrotron radiation for imaging at higher and higher spatial resolution. Improved emittance allows a beam of radiation to be focussed to smaller lateral dimensions, although the ultimate limit is determined by the wavelength of the radiation. In general, however, most (but not all) surface science techniques exploit radiation at somewhat lower energies, commonly more compatible with a reduced sampling depth in the selected technique. One limitation in exploiting improved emittance to produce beams that illuminate smaller and smaller spots on a sample is the problem of radiation damage, which generally scales with the flux density of the illuminating beam of radiation. This is a particular problem for organic molecular systems, both in the bulk (also confronted in transmission electron microscopy, where it is ameliorated by working at cryogenic temperatures) and also in the case of adsorption of such molecules on surfaces. Of course, in the case of diffraction studies of macromolecular crystals the need for radiation beams of small lateral extent is partly driven by the desire to study materials that can only be grown as very small crystals; in this case, one way to address the problem of radiation damage from extremely intense and narrow beams that can be produced by a free-electron laser (FEL) is to collect the data in extremely short times, at a timescale shorter than that of the crystal destruction. As yet, at least, such extreme situations are not confronted in surface science (although time-resolved pump–probe surface dynamics experiments may progress in this direction). The most obvious way of imaging, when a very narrow incident beam is available, is to collect an image by rastering the narrow beam over the surface, as in conventional scanning electron microscopy (SEM). The image intensity is then determined by some signal generated by the beam incident on the sample in each lateral position.

In general, one possible such signal is the transmitted intensity, but evidently transmission microscopy is not relevant for surface science. The absorption in the ultra-thin surface region of interest is masked by that from the much larger number of layers of substrate material. Scanned-beam surface imaging modes are therefore defined by the signal emitted from the surface as a result of the photon impact. In particular, detecting emitted electrons in an energy-selected mode leads to the scanned-beam form of photoelectron microscopy (SPEM), although detecting emitted electrons at other energies, or over a wide range of energies (partial or total yield), leads to microscopies based on the phenomena of XANES and X-ray magnetic circular dichroism (XMCD). All of these techniques are spectroscopic microscopies (and microspectroscopies) in which added information is provided by varying the detected electron energy or the incident photon energy at each position of the scanned incident photon beam. Notice that SPEM is fundamentally different in methodology from that of the full-field imaging photoelectron emission microscopy (PEEM) technique introduced in chapter 2 and described more fully in section 5.2.2, although both techniques provide images of a surface based on the intensity of photoelectron emission as a function of lateral position. However, PEEM achieves this using an incident photon beam that illuminates a (larger) fixed area of the surface, the spatial resolution of imaging within this illuminated area being achieved by the electron optics through which the emitted photoelectrons pass.

5.2 Photoelectron imaging

5.2.1 Scanning photoelectron microscopy

The key requirement for SPEM is a means of focussing the incident radiation to a small 'spot', generally of sub-micron lateral dimensions. A general survey of focussing optics is given in chapter 2, while figure 5.1 shows schematic diagrams of three relevant methods for SPEM. Notice that the Schwarzchild objective uses

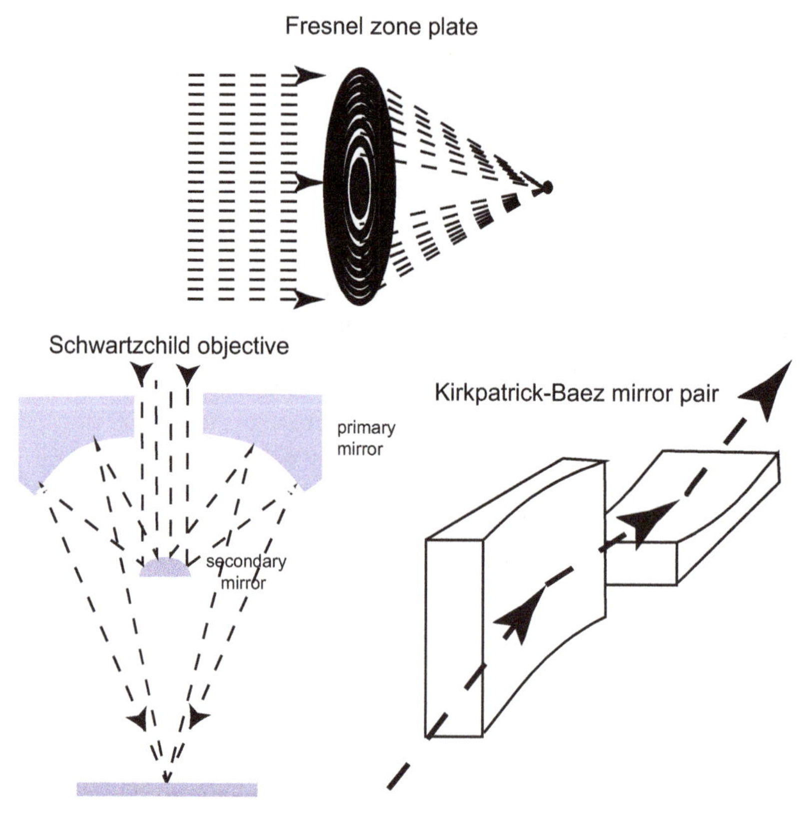

Figure 5.1. Schematic representations of three alternative incident beam focussing optics for use in SPEM. Note that the Schwartzchild objective can only be used in the soft X-ray energy range with interferometric multilayer coatings on the mirrors.

normal incidence reflections, which would thus be normally regarded as only relevant to use at long wavelengths close to those of visible light. However, by applying a multilayer coating onto the mirrors, such that one obtains coherent interference in the weak reflections of successive layers, significant reflectivity can be achieved at much shorter wavelengths, albeit tuned to a single value. A more conventional approach to focussing with reflective mirrors at shorter wavelengths is the use of a Kirkpatrick–Baez pair of spherical mirrors at grazing incidence, achieving separate focussing in the meridional and sagittal planes to overcome the strong astigmatism of a spherical mirror used at grazing incidence, as described in chapter 2.

Most widely used for SPEM instruments at synchrotron radiation sources are Fresnel zone plates (FZPs), the main features (and constraints) of which are discussed in chapter 2. While focussing down to lateral dimensions of a few tens of nanometres is possible in these instruments, with a typical working resolution of ~100 nm, the focal length that defines the working distance from the sample is only ~5–15 mm. Notice, too, that the focal length of FZPs is strongly dependent on the wavelength, so scanning the wavelength (as required in XANES) requires exceptional control of the position of the FZP relative to the sample.

SPEM images are obtained by recording some part of the photoemission spectrum at each position of the focussed beam on the surface, building up the image pixel by pixel. While the emission of interest is commonly a core-level peak providing elemental and chemical information on the lateral distribution of surface species, some applications also follow the intensity of ARPES features to map variations in electronic structure. Scanning of the relative beam position is commonly achieved by mechanical movement of the sample. In ARPES studies of the electronic structure of materials, an increasingly common problem is the challenge of growing sufficiently large single crystals of increasingly exotic materials, so having a highly focussed incident beam allows one to investigate small crystals, in some cases within a polycrystalline sample. Scanned images help to locate such small areas of interest.

An example of an application of SPEM, using an instrument installed on the ELETTRA facility in Trieste that uses FZP focussing, is shown in figure 5.2. The objective of this investigation, by Winkler *et al* (2021), was to determine the surface orientation dependence of the oxidation of rhodium. The sample comprised a polycrystalline Rh foil, the orientation of the individual crystallite surfaces being determined by electron back-scattering diffraction (EBSD), as shown in the image of figure 5.2(a) and the accompanying list of orientations in figure 5.2(b). Figures 5.2(c) and (d) show the Rh 3d X-ray photoelectron (XP) spectra from two of these crystal surfaces following exposure to 2.5×10^{-4} Torr of oxygen for 90 min at a temperature of 623 K, using an incident beam diameter of 0.13 μm. Also shown in these panels are representations of the structure of the associated clean-metal and oxidised surfaces. Fits to the XP spectra identify components due to metallic Rh, oxide (RhOx) and an interface state. Figure 5.2(e) shows a SPEM map of the relative intensities of the oxidic/metallic Rh three-dimensional components. Figures 5.2(f) and (g) show plots of this relative oxidic component as a function of step edge

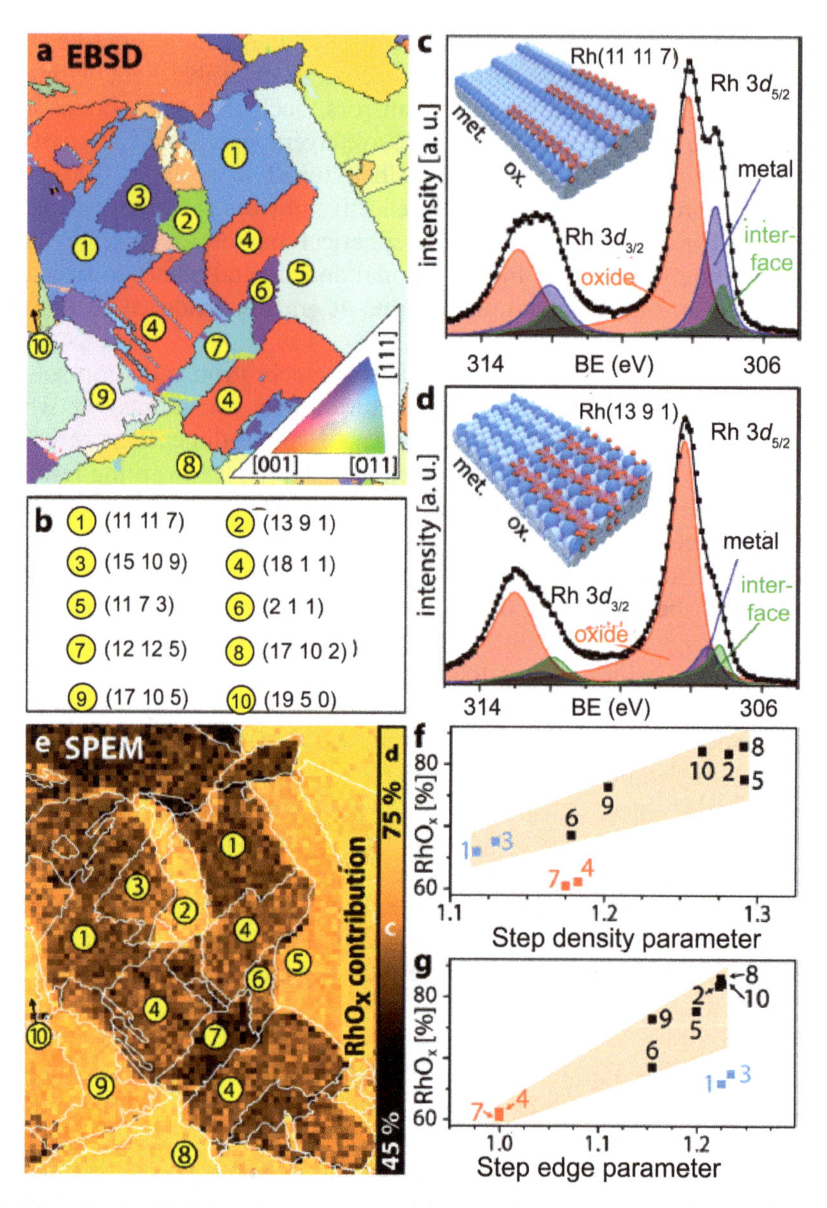

Figure 5.2. Results of a SPEM investigation of the surface orientation dependence of Rh oxidation. Panel (a) shows a EBSD map (512 μm × 600 μm) of the polycrystalline Rh sample, panel (b) lists the surface orientations of individual crystallites. Panels (c) and (d) show Rh three-dimensional XP spectra from two of these crystallites with the different chemical states identified, together with representations of the surface structures. Panel (e) shows a SPEM image of the surface based on the relative intensity of the oxidic Rh three-dimensional component. Panels (f) and (g) show plots of the relative oxidic components as a function of step edge density and step edge parameters. Reprinted by permission from Springer Nature, Winkler *et al* (2021), copyright (2021).

density and step edge parameters (as defined in the original paper) of the differently oriented surfaces.

Several examples of SPEM applications to investigate spatial variations of electronic structure have been concerned with the growth of graphene films on various substrates. Figure 5.3 shows an example of the results of one of these studies of graphene growth, with coexisting regions of graphene thickness from one to seven monolayers, on the C-face of SiC by Johansson *et al* (2014) using FPZ focussing at the ANTARES beamline of SOLEIL. Figure 5.3(a) shows a false-colour SPEM image of the spatial variation of the ARPES intensity, integrated over the range of initial states from 0.0–0.5 eV below the Fermi level, recorded with a photon energy of 100 eV in an emission direction corresponding to emission from the $\overline{K}$-point of the Brillouin zone. The different coloured regions correspond to graphene films of different thicknesses.

ARPES E–$k_{||}$ maps from four different areas, labelled (B) to (E), recorded with a spatial resolution of 120 nm, are shown in figures 5.3(b) to (e), and clearly show different numbers of linearly dispersing bands. The results of density functional

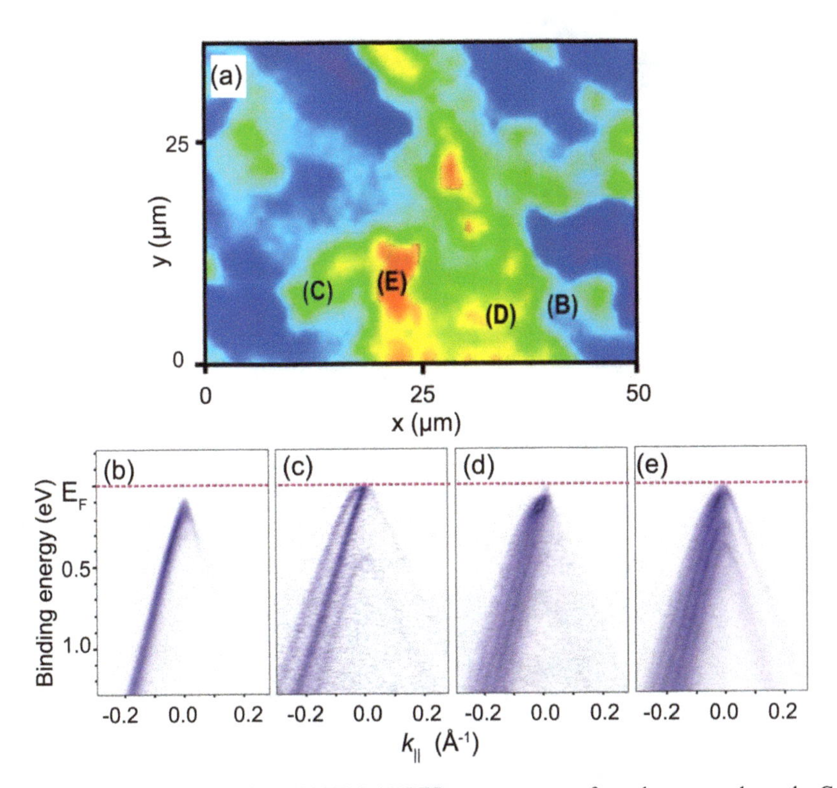

Figure 5.3. Results of the application of SPEM ARPES measurements of graphene growth on the C-face of SiC. Panel (a) shows a false-colour image of the intensity variation of the ARPES signal, integrated over the initial-state energy range 0.00–0.5 eV below the Fermi level, recorded at an emission angle corresponding to the $\overline{K}$-point of the Brillouin zone. Panels (b)–(e) show the E–$k_{||}$ maps obtained from the nano-ARPES recorded from regions (B)–(E) of the image in panel (a). Reprinted by permission from Springer Nature, Johansson *et al* (2014), copyright (2014).

theory calculations presented in this paper for different multilayer films of graphene in Bernal (AB) stacking are shown to be consistent with the observed multiple E–$k_{||}$ bands.

Figure 5.4 shows some results from a very different study of deposited graphene on a surface, namely an investigation of the ability of graphene on copper to impede oxidation of the Cu surface in a partial pressure of oxygen at a temperature of 623 K. In particular, these results, from an investigation by Amati *et al* (2023), performed SPEM at 'near ambient' pressures (NAP) to follow in operando a chemical reaction, in this case of corrosion (oxidation). On the left in figure 5.4(a) is

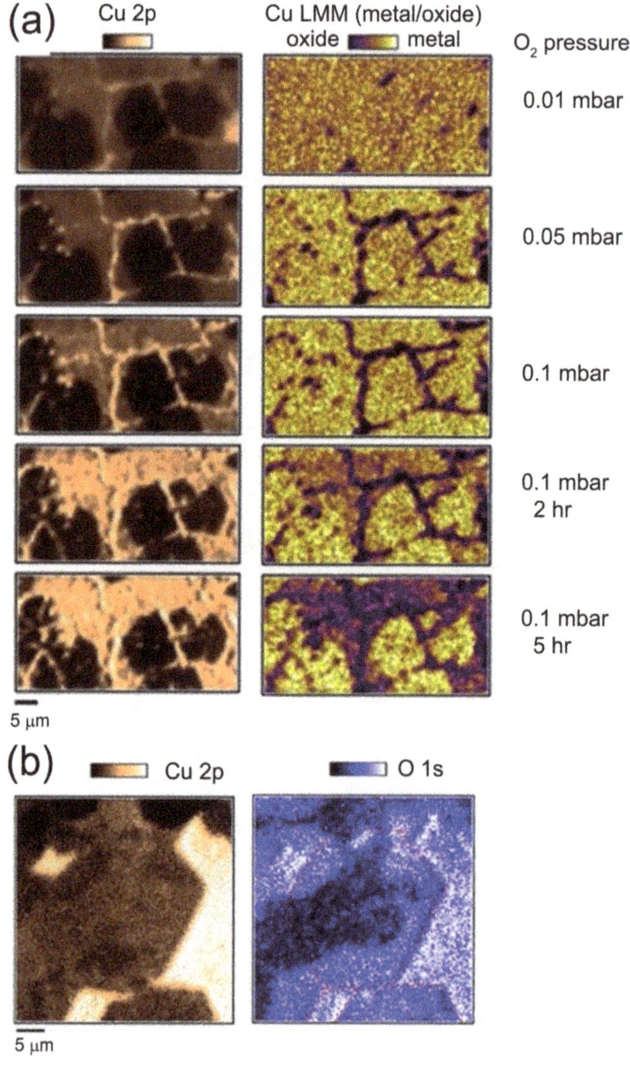

Figure 5.4. SPEM images recorded from a Cu surface covered with graphene flakes and exposure to different oxygen partial pressures for different times at a temperature of 623 K. Reprinted from Amati *et al* (2023), copyright (2023), with permission from Elsevier.

shown a series of SPEM images recorded with the Cu 2p photoemission signal under different O_2 pressures and after extended exposure to the highest pressure. The dark patches in the images correspond to the location of ~15 μm graphene flakes (which attenuate the Cu 2p emission intensity), the darkest patches being due to graphene bilayers while the less dark regions are covered with graphene monolayers.

The oxidation of the copper is monitored by the Cu LMM Auger signal; this spectrum contains two distinct peaks separated by ~2 eV, associated with metallic Cu and Cu_2O, respectively. The images on the right of figure 5.4(a) are formed by the relative intensity of the metallic and oxidic components, so oxidised regions of the surface show up as dark; these clearly correspond to the spaces between the graphene-covered areas. The oxidation of the copper does proceed with increasing oxygen pressure and extended exposure time, but is clearly retarded in the graphene-covered areas. Figure 5.4(b) compares Cu 2p and O 1s magnified SPEM images from one area. As in the images in figure 5.4(a), the Cu 2p signal is highest in the areas not covered by the graphene, but is also slightly higher in the areas under the centre of the central graphene flake than near its edges. The complementary O 1s image also shows its weakest intensity in the centre of this flake, but slightly higher intensity near its edges. This can be attributed to intercalation of oxygen under the graphene flakes from the oxidised areas, apparently a precursor to the complete oxidation of the copper.

5.2.2 Full-field photoelectron emission microscopy

A quite different approach to photoelectron microscopy is to illuminate a larger area of the surface of interest with photons and use electron optics to project an image of the spatial variation of the photoelectron emission within the illuminated area. The earliest PEEM instruments exploiting this idea, in the 1930s, were used with laboratory photon sources of ultraviolet (UV) radiation with photon energies ~4–6 eV, just sufficient to exceed the work function threshold for photoemission. As such, the resulting images were effectively maps of the spatial variation of the work function of the surface. A commercial instrument, the Balzers Metioscope, became available based on this approach and, as its name implies, was used mainly in metallurgical applications (even long after this instrument ceased to be manufactured; e.g., Hammond and Imam 1991), but this did not operate in UHV, thereby making it inappropriate for true surface science studies. Somewhat more recently, Engel *et al* (1991) developed a bolt-on instrument of this type that was fully compatible with modern UHV surface science chambers; figure 5.5 shows a simplified schematic of this instrument.

A key feature of this design, like other PEEM instruments, is the high voltage difference between the sample and first electron lens component, which strongly accelerates the photoelectrons into the detector. In this design the sample forms the cathode of a cathode lens system, a characteristic of most PEEM instruments. A brief but nevertheless detailed history of the development of PEEM instruments is presented in the paper by Bauer (2012a). While UHV PEEM instruments mainly used with UV radiation for work function images have proved valuable in a number

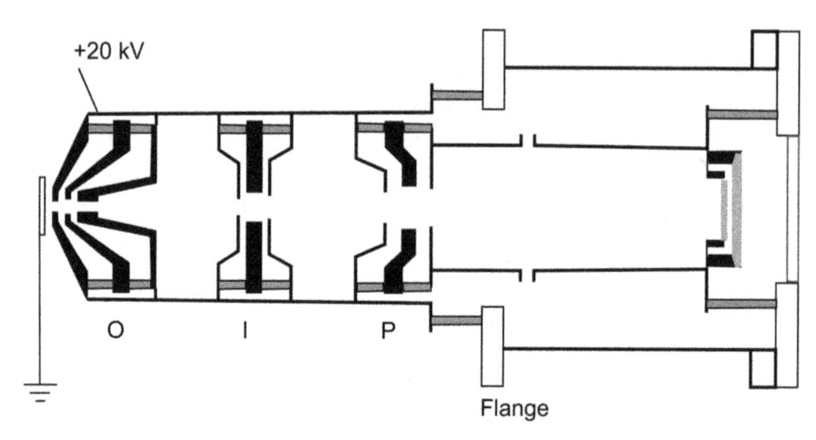

Figure 5.5. Simplified schematic diagram of the bolt-on PEEM instrument of Engel *et al* (1991) showing the flange used to attach the instrument to a UHV surface science chamber. Note the large voltage difference between the sample and the first lens element. Reprinted from Engel *et al* (1991), copyright (1991), with permission from Elsevier.

of static and dynamic surface studies (the SPEM investigation of Rh oxidation described in the previous section also used UV-PEEM to follow reduction of oxidised surfaces with hydrogen) they accrue no benefit from the use of synchrotron radiation and therefore lie outside the scope of this book. However, key aspects of the instrument designs are relevant to the PEEM studies using higher-energy synchrotron radiation soft X-rays (often referred to as XPEEM).

In fact, the first usage of synchrotron radiation for PEEM experiments appears to have been by Tonner and Harp (1988), using a simple instrument similar in concept to that reported by Engel *et al* (1991; see figure 5.5) with photon energies up to 160 eV. Of course, the potential benefit of using these higher photon energies is that they allow photoemission from core levels with energies characteristic of the elemental species. However, in this simple microscope design there is no energy filtering of the emitted electrons, so a very wide range of photoelectrons (and secondary and inelastically scattered electrons) contribute to the image. However, Tonner and Harp (1989) showed that by taking the difference between two images recorded using photon energies just below and just above a core-level photo-ionisation threshold, a more element-specific image could be obtained. In effect, therefore, they demonstrated the value of spatially dependent XANES. Of course, an instrument with true 'band-pass' energy filtering is to be greatly preferred, and this is a characteristic of modern PEEM instruments used at synchrotron radiation sources, although many synchrotron radiation PEEM studies do not exploit this energy-filtering capability.

In particular, one widely used instrument for XPEEM at synchrotron radiation beamlines is actually the low-energy electron microscope (LEEM) as introduced in section 3.5 in the context of ARPES electronic band mapping. Figure 5.6 shows a simplified schematic of the LEEM instrument pioneered by Bauer in the 1960s, now available commercially in various forms from the Elemitec company in Germany; Bauer (2012b) provides a historical survey of this development. The underlying idea

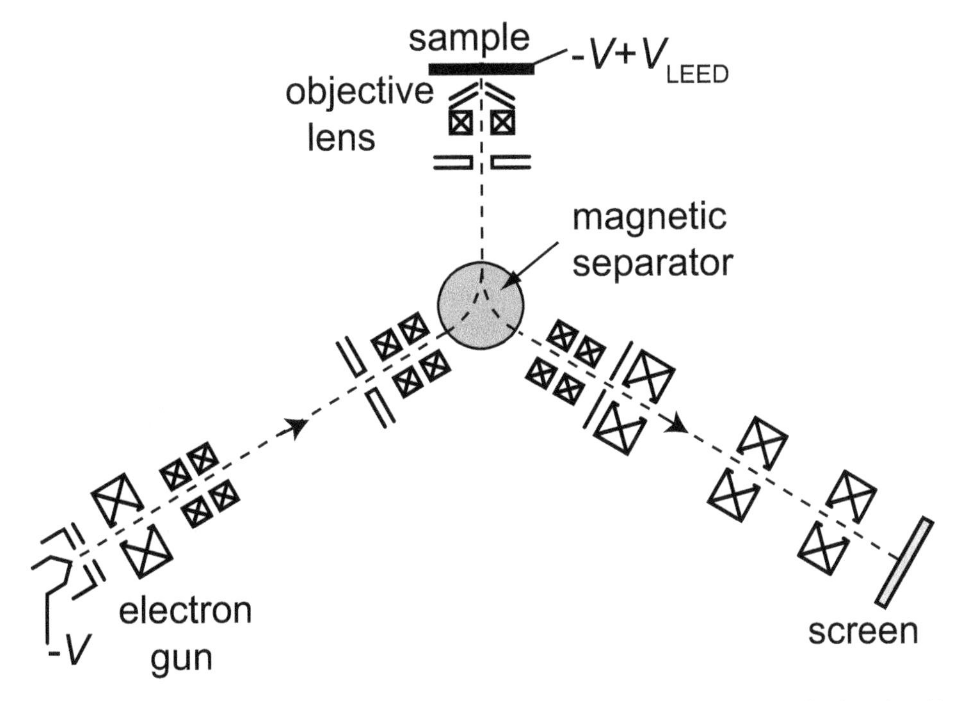

Figure 5.6. Simplified schematic diagram of a LEEM instrument (also used for XPEEM) as first introduced by Bauer in the 1960s.

of a LEEM is to provide a method of imaging a surface, based on low-energy back-scattering diffraction, analogous to the transmission electron microscopy (TEM) used with great success to investigate the structure of crystalline solids based on high-energy forward-scattering electron diffraction. Low-energy electron diffraction (LEED) is well-known to provide surface-specific structural information and can be used with the aid of multiple scattering simulations to provide quantitative atomic-scale determinations of the surface structure. However, the electron energy range used in LEED, from a few eV to a few hundreds of eV, is too low for electron optics to provide high-resolution images due to the combined effects of space charge and stray magnetic and electrostatic fields. To overcome this problem in a practical LEEM instrument, the electron optics for incident beam delivery and image formation from the diffracted beams operate at much higher electron energies (~15 keV), using components designed for TEM operation. By applying a large repulsive potential to the sample, the electrons actually scatter from the sample at LEED energies.

The elastically scattered (diffracted) electrons are then reaccelerated by the same objective ('cathode lens') and deflected in the opposite direction by the magnetic field that separates the incident and diffracted electrons, which then pass into an electron imaging column, also operating at the same high energy as the electron gun. This type of instrument can operate not only as a LEEM but also as a PEEM. If the electron gun is switched off, and the sample illuminated with photons, the emitted photoelectrons, like the diffracted low-energy electrons from electron incidence, are accelerated in the cathode lens and pass through the imaging optics to produce a PEEM image. Moreover, as in a TEM, modifications to the operating mode as a LEEM allows one to observe either the diffraction pattern or the real-space image. Operating with incident photons instead of electrons, the operating mode to display a diffraction pattern now displays the angular distribution of the photoemission. The full benefit of a PEEM instrument, however, still requires some kind of energy filter, and one solution is to place a concentric hemispherical electron energy analyser (CHA; see figure 3.18) at the exit of the imaging column; this approach has been implemented and is widely used in combination with the LEED/PEEM instruments of figure 5.6. Figure 5.7 shows schematically the different LEEM/PEEM operation modes leading in PEEM to (a) energy-filtered PEEM images, (b) angular distribution of the photoemission at a specific electron energy and (c) XP spectra. Note that the resolving power ($E/\Delta E$, where E is the energy of the electrons passing through the analyser) of a CHA is determined by the ratio of the mean radius of the hemispheres and the exit slit. To achieve an acceptable energy resolution of a few tenths of an eV is therefore necessary to use retarding electron optics between the high-energy (up to tens of keV) optical column and the CHA to lower the pass energy through the CHA to only ~100 eV. This combined LEEM/PEEM + CHA instrument has been referred to as a spectroscopic photoemission and low-energy electron microscope (or SPELEEM; e.g., Menteş and Locatelli 2012).

An alternative design of LEEM uses a deflection angle of 90° between the incident and diffracted electrons, leading to a geometry in which the axes of the electron gun and the imaging optics are colinear (Tromp *et al* 2010). This instrument is also available commercially from the SPECS company, also in Germany. As in the design of figure 5.6, operating this instrument in the XPEEM mode involves

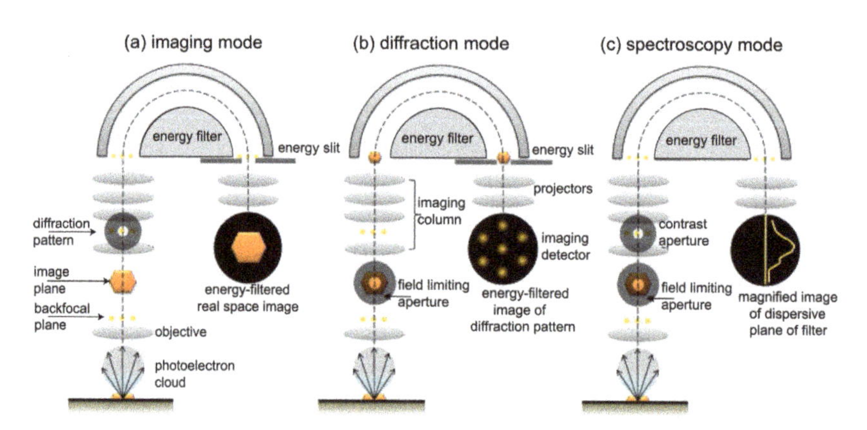

Figure 5.7. Schematic diagrams showing three modes of operation of the LEEM/PEEM instrument of figure 5.6 combined with energy analysis/filtering using a CHA 'energy filter'. Panel (a) shows the mode of operation generating a LEEM image or an energy-filtered PEEM image. Panel (b) shows the mode generating a diffraction pattern in LEEM or an angular distribution of photoemitted electrons in PEEM. Panel (c) shows the operating mode leading to a XP spectrum in PEEM. Reprinted from Menteş and Locatelli (2012), copyright (2012), with permission from Elsevier.

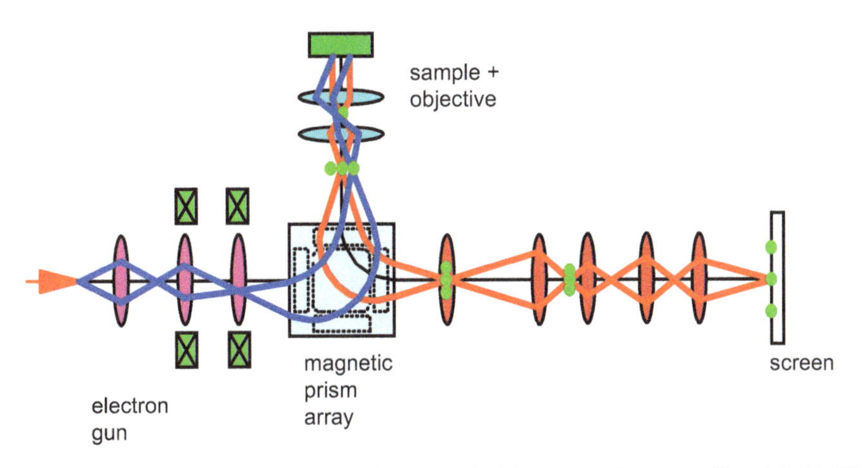

Figure 5.8. Schematic diagram showing the electron optical layout of the energy-filtered LEEM/PEEM instrument of Tromp *et al*. Reproduced from Tromp *et al* (2009).

switching off the electron gun and illuminating the sample with photons to stimulate photoemission, the photoemitted electrons being accelerated into the imaging optics. Figure 5.8 shows the optical layout of this instrument. Superficially, the only difference from the design of figure 5.4 is that the magnet structure separating the incident and diffracted electrons in the LEEM mode produces a deflection of 90° instead of 60°. However, the design of this magnet structure has further important properties. Firstly, it produces focussing of the diffracted or emitted electrons, but in addition the fact that this focussing is achromatic means that by insertion of a suitable aperture in the imaging column the electrons that pass on are energy filtered (Tromp *et al* 2009). Energy filtering thus occurs where the electrons are at the high energy of the microscope optics, with no need for the strong retardation before the electrons are passed into the CHA as in the design of figure 5.7.

One important improvement in both of these LEEM designs—and also at least one other PEEM instrument, spectromicroscopy with aberration correction for many relevant techniques, or SMART, installed on the BESSY synchrotron radiation source—has been the introduction of additional electron optics for aberration correction (Tromp *et al* 2010, 2013, Schmidt *et al* 2013), following the same development that has had a major impact in improving the spatial resolution in TEM instruments. This development has resulted in a significant improvement of the spatial resolution that can be achieved in LEEM mode in both designs of LEEM/PEEM microscopes to approximately 2 nm. This same improvement is implemented when these instruments are operated in the PEEM mode, although the achievable spatial resolution is roughly an order of magnitude worse. The reason for this appears to be the time structure of synchrotron radiation. As described in chapter 2, for most experiments in surface science, including (in general) photoemission, the time structure of synchrotron radiation can be regarded as quasi-continuous, although in reality it comprises short (~tens of picosecond) pulses separated by gaps of ~2 ns. The instantaneous photon flux in each of these pulses is thus about two orders of magnitude larger than the time-averaged flux, so this also applies to the rate of photoelectron emission. As a result of this, the resulting photoelectron cloud in front of the surface can experience 'space charge' effects (Schmidt *et al* 2013), due to electron–electron repulsion in the photoelectrons closest to the surface. It is this effect that is believed to degrade the spatial resolution in PEEM, even if the instrument is fitted with aberration correction (indeed, space-charge effects can also arise in the aberration correction optics). Notice that this space-charge effect can be even more important in the use of photoemission for a FEL, which produces even more intense and shorter pulses of incident radiation (e.g., Hellmann *et al* 2012). Niu *et al* (2023) have recently presented a detailed analysis of the spatial resolution achievable in an aberration-corrected Elmitec microscope installed on the MAXIV synchrotron radiation facility. They show that the effects of space charge can be significantly reduced by attenuating the incident photon flux by detuning the undulator; they also show the effects of modifying the mode of use of the electron optics to reduce the photoelectron current within parts of the microscope. Some examples are shown in figure 5.9.

Figure 5.10 shows results of a PEEM investigations of two-dimensional lead halide perovskite single crystals by Liang *et al* (2023). As indicated in the figure, three different samples of the general formula $(BA)_2(MA)_{n-1}Pb_nI_{3n+1}$ were

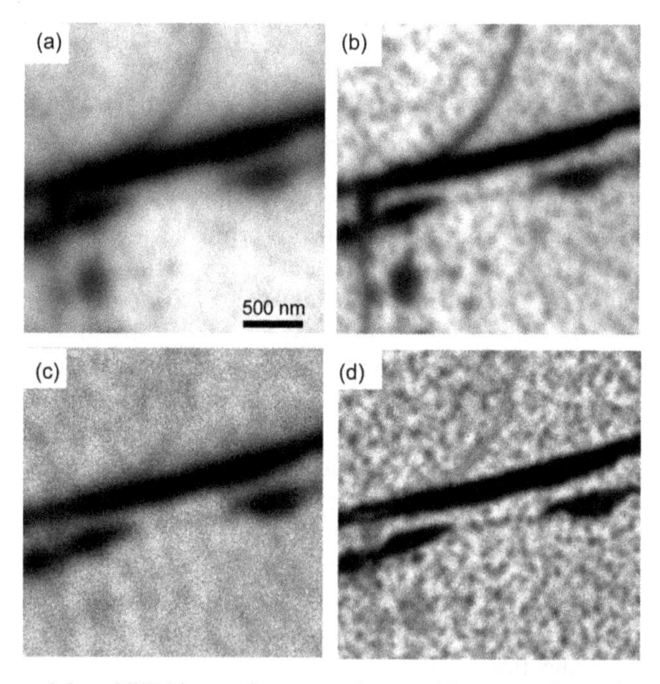

Figure 5.9. 2.5 μm × 2.5 μm PEEM images from a sample comprising a monolayer of graphene, with some bilayer areas, on SiC(0001). Panels (a) and (b) are recorded in the secondary electron mode and show the influence of reducing the incident photon flux from (a) 4.2×10^{13} photons/s to (b) 1.4×10^{13} photons/s. Panels (c) and (d) are energy-filter images recorded at the kinetic energy corresponding to Si 2p core-level emission at a photon energy of 150 eV recorded with the high flux of (a). Panel (d) shows the effect of introducing certain apertures in the microscope imaging column. Reproduced from Niu *et al* (2023). CC BY 4.0.

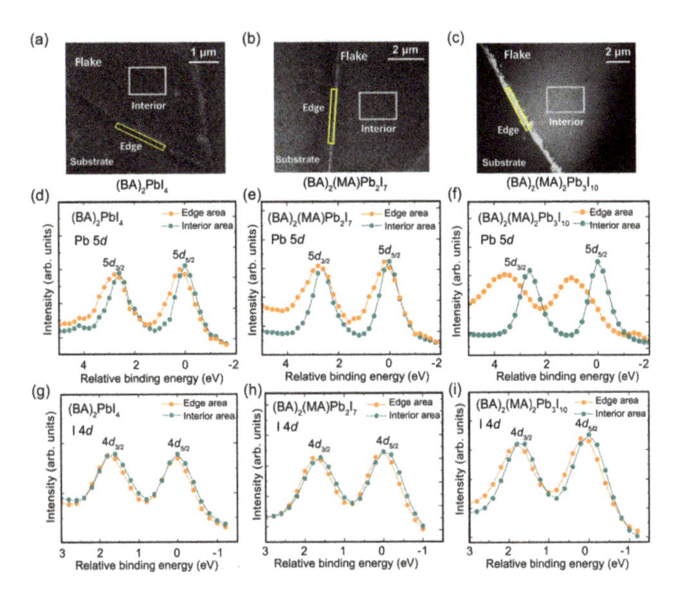

Figure 5.10. Results of a PEEM study of the spatial variation of the electronic structure of three different two-dimensional lead halide perovskite crystals. Panels (a)–(c) show PEEM images recorded with Pb $5d_{5/2}$ emission. The superimposed rectangles define areas at the crystal edges and interiors, over which Pb 5d and I 4d XP spectra were recorded; these are shown in panels (d)–(i). Liang *et al* (2023) John Wiley & Sons.

investigated, where BA is butylamine and MA is methylamine. These are quantum-well materials in which n is the number of octahedra layers that determine the quantum-well thickness. They are of interest for their optoelectronic properties and potential applications in, for example, solar cells and light-emitting diodes. Earlier investigations had indicated that so-called lower edge states occur at crystal edges and that these may facilitate dissociation of photogenerated excitons, further improving their application of optoelectronic devices. The objective of the PEEM study was to spatially resolve the electronic properties of these samples.

XP spectra shown in figures 5.10(d)–(i), recorded from regions close to the edge and in the interior of the crystals (identified in the PEEM images of figures 5.10(a)–(c)), show that there are significant differences in the photoelectron binding energies of the Pb 5d peaks between these different areas, although any differences are much smaller in the I 4d peaks. In the case of the crystals with $n = 1$ and 2 the energy shifts are only about 0.2 eV, and in the $n = 3$ case the shift is approximately 1 eV. The authors of this study rationalise this difference as due to strain relief at the crystal edges.

5.2.3 SPEM vs PEEM?

With two alternative methods of photoelectron imaging, it is of interest to compare the various aspects of SPEM and PEEM, although the fact that there are different implementations of both methods at different facilities makes it difficult to make broad generalisations. The fact that images in SPEM are achieved using mechanical rastering of the sample position, whereas PEEM creates the whole image area simultaneously tends to favour PEEM in terms of speed, although to obtain PEEM images of the highest resolution can require summation of multiple images to achieve good image quality. Similarly, a comparison of the spatial resolution that can be achieved depends on whether one judges typical performance or the ultimate achievable conditions. In PEEM the limiting factor for a resolution of less than a few tens of nanometres seems to be space-charge effects, but with lower photon flux and optimisation of the microscope operating conditions a resolution of ~10 nm or even slightly less seems to be possible. In SPEM the spatial resolution is determined only by the lateral dimensions of the incident photon beam. Claims have been made of possible values of less than 10 nm, but practical demonstrations are limited to a few tens of nanometres, also typical of claims for PEEM instruments. In practice, many studies with both methods concentrate on using sub-micron resolution. Finally, one may compare the achievable energy resolution (in both images and spectral recording). Here, the SPEM technique, constrained only by the typical performance of CHA instruments working with relatively low photoelectron energies of no more than a few tens of eV, can deliver resolution in the meV (or at least tens of meV) range, whereas PEEM instruments more commonly operate with resolutions of a few hundred(s) of meV.

One possible difference in the applicability of the SPEM and PEEM techniques is to NAP investigations. In PEEM instruments there is a very large voltage difference between the sample and the first lens element, which may be separated by only ~2 mm or less, leading to a local field $\sim 10^7$ V m^{-1}. To avoid arcing the sample surface the roughness needs to be low, but another consequence is that this design would seem to preclude NAP experiments, as this field would lead to a gas discharge breakdown. However, applying a modification of the extraction optics in the instrument of figure 5.8 such that the first lens ('nozzle') element is at a voltage much closer to that of the sample, while the strong acceleration is performed in the next lens element, means that NAP-PEEM can be performed in this type of instrument (Ning *et al* 2019), and indeed SPECS offer a modified version of their commercial PEEM for this purpose. Of course, as in NAP-XPS (section 3.2.1), a special cell and differential pumping is also required. Figure 5.11 shows two images of graphene on Ru(0001) recorded on this modified instrument using a mercury light source to produce UV-PEEM images with a spatial resolution of ~20 nm in a partial pressure of 0.1 mbar N_2 (as shown by the inset line scans); the resolution was found to degrade to ~100 nm when the pressure was raised to 1.2 mbar.

5.3 XANES imaging

While energy filtering in a PEEM is necessary to provide element-specific and chemical-state-specific photoelectron images, and it is also essential for the ARPES mode of operation of a PEEM instrument, it transpires that many practical applications of PEEM are of XANES imaging for which photoelectron energy filtering is not required, as shown by the early experiments of Tonner and Harp (1988, 1989). Recall that a standard method of measuring XANES is by monitoring the total electron yield as a function of photon energy. Chemical contrast can

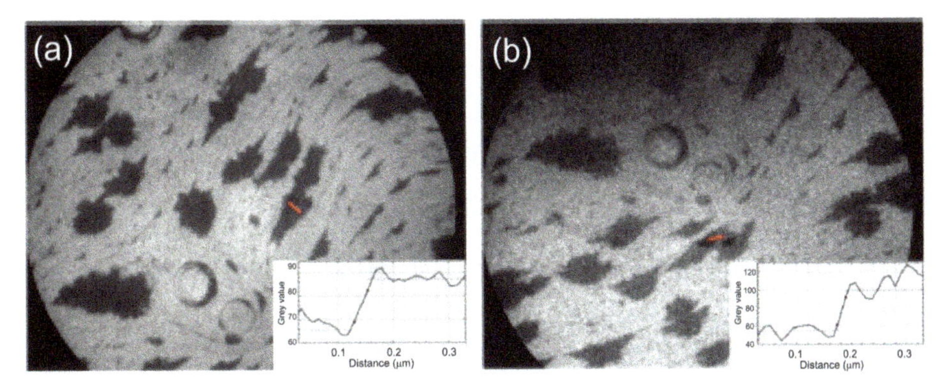

Figure 5.11. UV-PEEM images of graphene flakes on Ru(0001) taken in (a) UHV and (b) 0.1 mbar N_2. The field of view is 14 μm. The inset line profiles, corresponding to the red lines superimposed on the images, indicate a spatial resolution of 20 nm. Reprinted from Ning *et al* (2019), copyright (2019), with permission from Elsevier.

therefore be achieved by recording the difference between images obtained at photon energies just below and just above core-level photoionisation energies. A particularly important form of XANES imaging, however, is that afforded by exploiting variable polarisation modes to obtain spatially resolved XMLD and XMCD data (see section 3.7) to study domain structures in ferromagnets, antiferromagnets and ferroelectrics (e.g., Schneider and Schönhense 2002). An early example of this is the XMLD imaging study of antiferromagnetic NiO(100) thin films grown on MgO (100) by Stöhr *et al* (1999), presenting the ratio of PEEM images recorded at two different photon energies corresponding to the two main peaks in the L_2-edge XANES spectrum. Their images, recorded at room temperature, showed contrast in defect structure that disappeared as the sample temperature was increased, leading to an identification of a reduced Néel temperature of the defect structure relative to that of the bulk.

Using secondary electron imaging in PEEM it is important to be able to distinguish between image contrast due to composition, to topography and—in the case of XMCD and X-ray linear dichroism (XLD) imaging—due to magnetic or ferroelectric domain structure. Appropriate processing of the collected raw images can achieve this. Figure 5.12 shows an example of how this was achieved by Schneider and Schönhense (2002) to image the XMCD asymmetry in permalloy microstructures.

The intensity of individual pixels in images recorded with opposite incident radiation helicities are summed, subtracted and ratioed to produce a final asymmetry image of the local XMCD values. The change in sign of the XMCD signal at the L_2 and L_3 images can also be used to enhance the image contrast.

The angular dependence of the magnetisation vector, $\boldsymbol{M}(x,y)$, and the radiation helicity, ζ, influences the image contrast of magnetic domain structures allowing domain magnetisation directions to be determined. Specifically, the contrast $A \sim \boldsymbol{M}.\zeta \sim A_0 \cos\phi$, where ϕ is the angle between these two vectors. Figure 5.13 shows

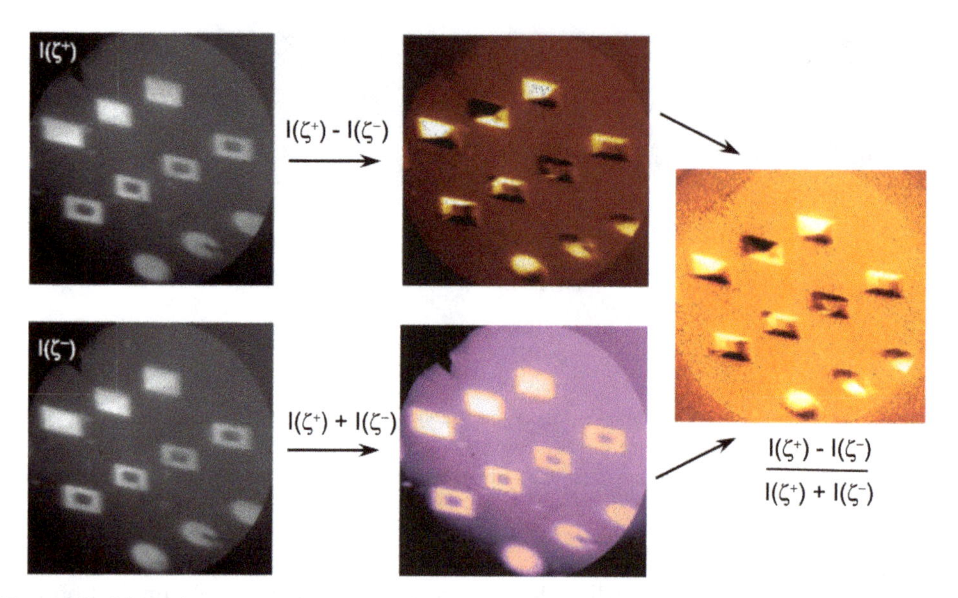

Figure 5.12. Magnetic contrast enhancement of XMCD images of 12 μm permalloy microstructures. On the left are individual images recorded at the Ni L_3 edge using the opposite incident radiation helicities, $I(\zeta^+)$ and $I(\zeta^-)$. In the centre are shown the images prepared from the difference and sum of the results from the two helicities. These are used to calculate the asymmetry image shown on the right. Reproduced from Schneider and Schönhense (2002). © IOP Publishing Ltd. All rights reserved.

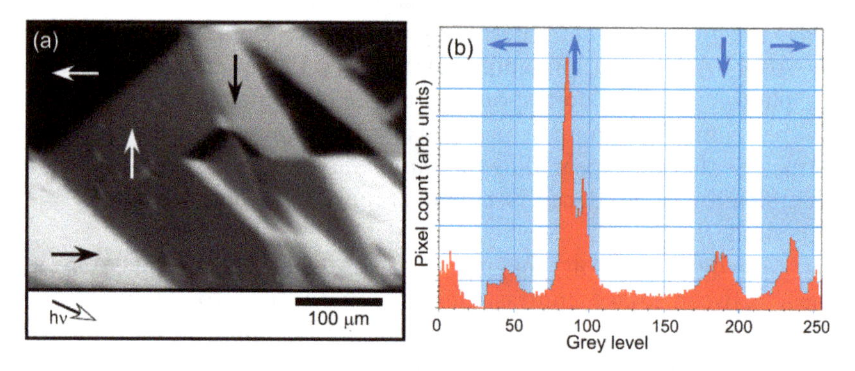

Figure 5.13. (a) XMCD-PEEM image of the magnetic domain pattern on a ferromagnetic Fe(001) whisker surface obtained at the Fe $L_{2,3}$ edge. The arrows denote the local orientation of the magnetisation vector. (b) Histogram of the grey levels in the image, revealing four maxima (shaded regions) corresponding to four distinct magnetisation directions. Reproduced from Schneider and Schönhense (2002). © IOP Publishing Ltd. All rights reserved.

an example of the use of this effect in imaging the magnetic domain pattern on a ferromagnetic Fe(001) whisker.

One example of the use of both XMCD-PEEM and XLD-PEEM to investigate magnetism and ferroelectricity is illustrated by a group of papers by Ghidini *et al* (2018, 2020, 2022), who studied a sample comprising a 10 × 10 array of 1 μm-diameter 10 nm-thick nickel (Ni) discs on a metallised 0.5 mm-thick BTO ($BaTiO_3$)

ferroelectric substrate, shown schematically in figure 5.14(a). The Ni discs were terminated with a 3 nm Cu protective cap. Figures 5.14(b) and (c) show 15 μm elemental Ni and Ti XANES images obtained by differences of images recorded at photon energies just below and just above Ni L_3 and Ti L_3 absorption edges. Figure 5.14(d) shows a composite of a 20 μm-diameter XMCD-PEEM image of the Ni discs combined with an XLD-PEEM image of the BTO substrate (the green

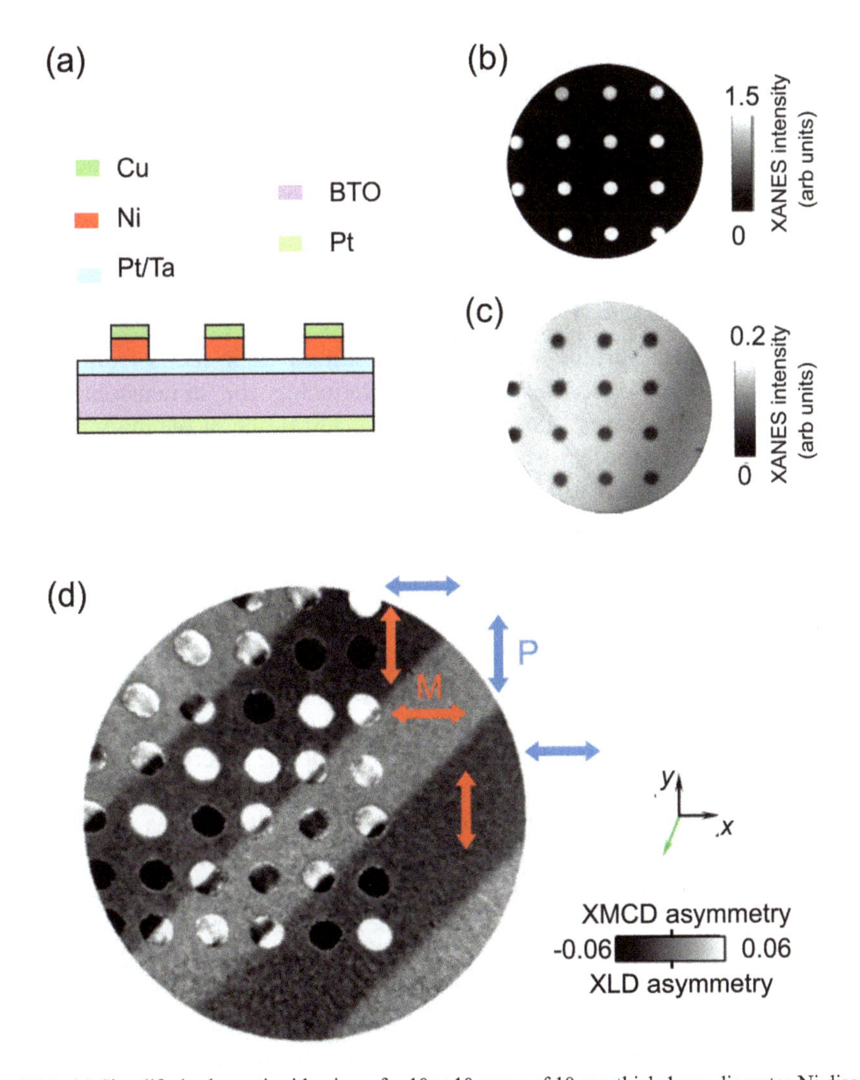

Figure 5.14. (a) Simplified schematic side view of a 10 × 10 array of 10 nm-thick 1 μm-diameter Ni discs on a BTO ($BaTiO_3$) substrate used in a combined XMCD-PEEM and XLD-PEEM series of experiments by Ghidini *et al.* Panels (b) and (c) show 15 μm Ni and Ti elemental XANES-PEEM images. Panel (d) shows a composite 20 μm-diameter MXCD-PEEM image of the Ni discs with an XLD-PEEM image of the BTO substrate. Further details described in the main text. Panels (a) and (b) reproduced from Ghidini *et al* (2020) with permission from the Royal Society of Chemistry. Panel (c) reproduced from Ghidini *et al* (2022). Panel (d) reproduced from Ghidini *et al* (2018).

arrow shows the plane projection of the incident beam direction). For each ferroelectric domain double-headed arrows show the orientation (without sign) of the polarisation **P** (blue) and local magnetisation **M** (red). The alternating dark and light stripes correspond to opposite ferroelectric polarisation directions of the BTO substrate. Ni discs that fall entirely within one type of ferroelectric stripe show high MXCD signals being entirely black or white, corresponding to the local magnetisation vector being antiparallel or parallel to the in-plane direction of the incident X-ray beam (green arrow). These discs are therefore inferred to be single domain. Ni discs that straddle two different ferroelectric polarisation directions show two different magnetic domain orientations, consistent with the fact that the size of the Ni discs exceeds the critical diameter for these Ni films. More detailed investigation of the MXCD-PEEM images of individual Ni discs under varying applied voltages to the BTO substrate revealed considerably more detail in the surface magnetism changes in this series of studies.

5.4 Coherent X-ray diffraction imaging of nanoparticles

In recent years many surface scientists have shifted their focus from essentially planar surfaces to nanoparticles, effectively reducing the dimensionality of the physics from two dimensions to one or zero. Of course, one of the main motivations for surface science has been to understand heterogeneous catalysis, and practical catalysts are always in the form of small particles, particularly to increase the active surface area of expensive materials. To what extent this chemistry is influenced when the small particles have nanometre dimensions, potentially changing their structure and electronic and chemical properties, is clearly a topic of interest. An ability to image such exceptionally small particles and understand their structure and internal behaviour is a goal that is being achieved with synchrotron radiation through what is commonly referred to as lens-less X-ray imaging. The underlying idea goes back again to the Abbé theory of the microscope. The first stage of creating and measuring a diffraction pattern is the first stage of the microscope, but creating an image can then be performed numerically directly from the diffraction pattern rather than by using a lens. A key requirement to exploit this idea successfully is a coherent source and, as described in section 2.4, synchrotron radiation is not intrinsically coherent in the same way as a conventional laser or FEL. However, by limiting the radiation to pass through a pinhole, a coherent radiation source can be created, albeit at reduced intensity. The simplest arrangement for this plane-wave coherent X-ray diffraction imaging (PCXDI) experiment with the pinhole source, the sample and a two-dimensional detector is shown in figure 5.15, albeit omitting an essential beam stop to block the direct beam from striking the detector.

Of course, the numerical procedure to construct the sample image from the recorded diffraction pattern must confront the phase problem of X-ray diffraction, namely that the image can be obtained from the Fourier transform of the diffracted amplitudes, but one can only measure the diffracted intensities; the phase information is lost. This problem is well-known in X-ray crystallography and has also been mentioned in the description of the surface X-ray diffraction (or SXRD) technique

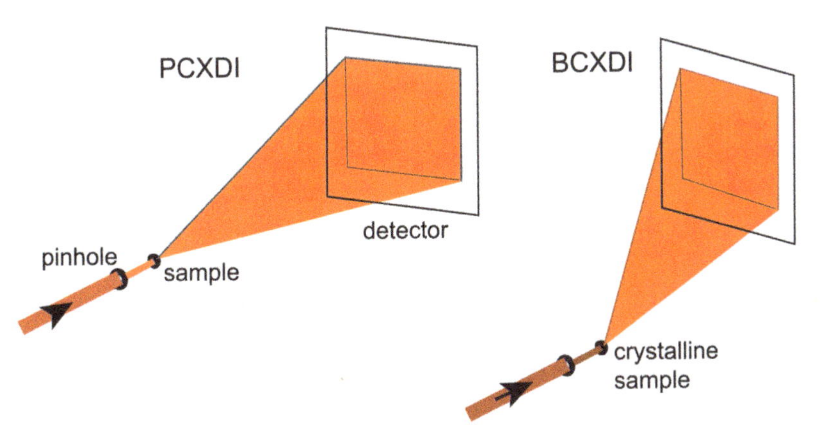

Figure 5.15. Simplified schematic diagram showing the key components of PCXDI and BCXDI experiments.

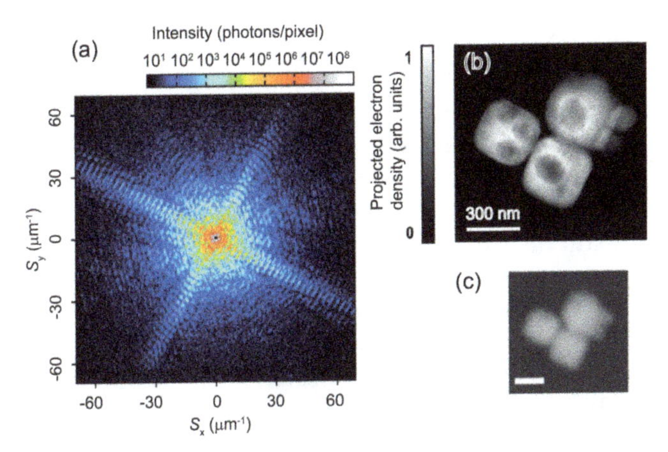

Figure 5.16. Results of a PCXDI study of dispersed 400 nm gold particles by Takayama *et al* (2018). Panel (a) shows the recorded diffraction pattern, (b) the recovered image. A SEM image of the same particles in shown in (c). Reproduced from Takayama *et al* (2018) with permission of the International Union of Crystallography, according to the terms and conditions of use of material published by the International Union of Crystallography.

(section 4.1) and in section 3.5.3 in the context of the photoemission orbital tomography technique. The general solution in all these cases involves an iterative procedure based on initial 'guessed' phases; these methods are well-established in X-ray structure determinations. A key requirement for their success is to have a sufficiently large database of measured intensities. In this respect, coherent X-ray diffraction imaging (CXDI) has the advantage over measurements of diffracted beams from a crystal structure in that the sampling in diffraction space is continuous rather than discrete, so it is possible to oversample the pattern relative to the spatial Nyquist frequency.

Figure 5.16 shows the results of a PCXDI study of 400 nm gold particles dispersed on a silicon nitride membrane by Takayama *et al* (2018). Figure 5.16(a) shows the recorded diffraction pattern, while figure 5.16(b) shows the PCXDI-derived image, which is compared with a conventional SEM image of the same sample. Of particular interest is the clear evidence of voids in the gold particles that can be seen in the PCXDI image (but not in the SEM image).

Dispersed gold particles of this type have been used as samples to explore the CXDI technique by a number of groups, and using a nanofocussed incident beam Schroer *et al* (2008) showed that using the coherent part of a ~100 nm-diameter beam it was possible to image a single gold particle smaller than 100 nm with a resolution of 5 nm. Using a modified form of CXDI, however, it is possible to image single particles from an illuminated area containing a distribution of particles. Notice that the PCXDI technique does not rely on the samples being crystalline. However, if they are crystalline, images of individual particles can be extracted from a distribution of particles, illuminating a wider area, but using the slightly different technique of Bragg CXDI (BCXDI; see also figure 5.15). If the many particles are randomly oriented the Bragg-diffracted beams emerge in different directions (as in a conventional powder X-ray diffraction experiment), so the diffraction patterns of single particles can be measured separately. A further important advantage of BCXDI is that because of the underlying use of Bragg crystalline diffraction, images can provide information on the internal modifications of the crystalline order within the particles. One example of this is the study of gold (Au) nanoparticles subject to Cu deposition and annealing by Xiong *et al* (2014). Figure 5.17 shows BCXDI Au (111) diffraction patterns from a single ~300 nm Au nanocrystal deposited on a silicon substrate, subjected to slow deposition of Cu. Cu and Au are known to form bulk-ordered alloy phases of stoichiometry Au_3Cu, CuAu and Cu_3Au, so the experiment aimed to provide information on the diffusion of the Cu into the original Au nanocrystal. The sample was held at a temperature of 300 °C to ensure that detectable diffusion effects could be monitored over a period of a few hours.

The diffraction pattern prior to the Cu deposition (figure 5.17(a)) is fairly symmetric and shows fringes that are most prominent in the facetted directions of the nanocrystal. After longer times the patterns become less symmetric, with distorted fringes attributed to inhomogeneous lattice distortions due to the diffusion of the Cu atoms into the Au crystal. The scale bars show that the maximum intensity decreases with time, and after 10 hr the sample showed powder diffraction rings consistent with Cu_3Au; the reduced intensity of the (Au(111)) diffraction pattern is consistent with the Au crystal volume shrinking as the different structure of the Cu_3Au phase is formed.

Phase and amplitude reconstructions from these diffraction patterns are shown in figure 5.18. The phase is a projection of the lattice distortion onto the (111) **G** vector (shown as an arrow in figure 5.18(b)). Initially, the nanocrystal is essentially strain-free, with little variation of phase in the nanocrystal; notice that outside the crystal the phase reconstruction is very noisy because the amplitude here is essentially zero. However, after 6 hr (figure 5.18(d)) there are three characteristic dipole-shaped features, two near the top and one on the left. Each shows a phase change of $-\pi$ to $+\pi$ over a small volume, indicating atom displacements equal to the Au(111) lattice spacing of 2.35 Å. These are attributed to dislocation loops formed by the insertion of a single plane of atoms over a small region between two lattice planes, with an

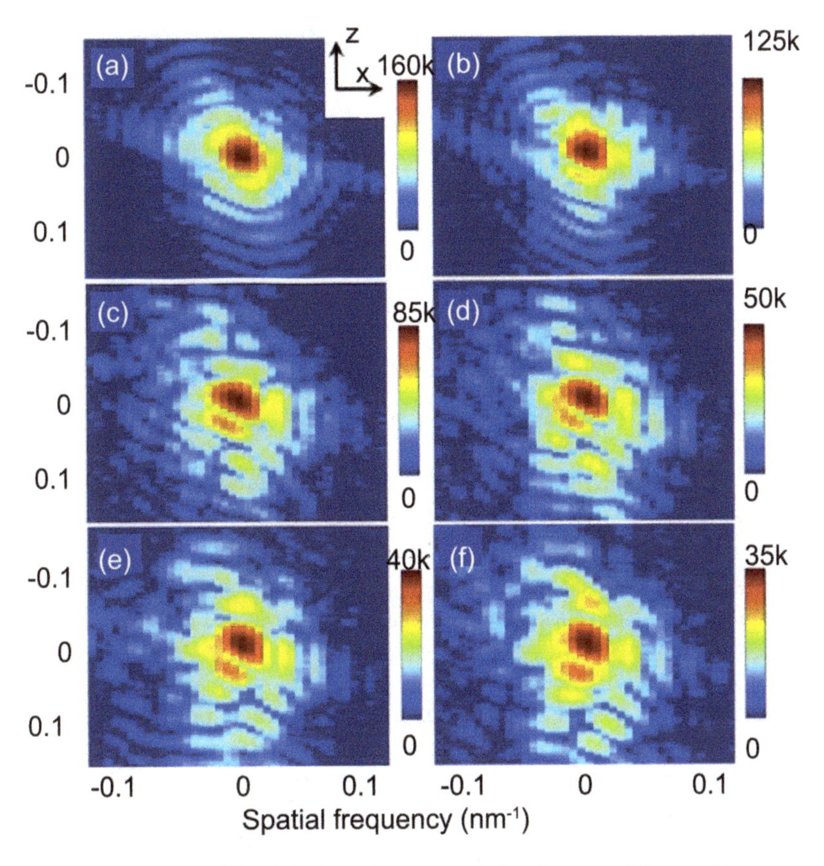

Figure 5.17. *X–Y* plane views of BCXDI diffraction patterns taken from a ~300 nm gold nanocrystal under continuous Cu deposition at the centre position of the Au(111) rocking curve. Times after Cu deposition was started are (a) 0 hr, (b) 2 hr, (c) 4 hr, (d) 6 hr, (e) 8 hr, and (f) 10 hr. Reproduced from Xiong *et al* (2014). CC BY 4.0.

associated strain field surrounding it. The amplitude reconstructions of figures 5.18(g)–(l) show that the (Au(111)) crystalline volume (the volume giving rise to the Au(111) Bragg peak) appears to shrink as diffusing Cu atoms modify the local Au crystalline order and 'channels' start to open up in the crystal—the authors suggest that these channels are probably not empty, but correspond to regions where the Cu atoms can diffuse and modify the structure.

The fact that Cu and Au form ordered bulk alloys clearly shows that Cu diffuses in bulk Au, and indeed bulk diffusion rates are well-known. However, the diffusion behaviour may be expected to differ in a nanocrystal, and modelling of the diffusion in this nanocrystal, based on a two-shell model, led the authors of this study to conclude that the diffusion coefficient in this experiment was 8.7×10^{-9} μm^2 s^{-1}, some two orders of magnitude larger than in bulk Au at 300 °C.

Notice that the phase-retrieval methods of CXDI can lead to ambiguities if the sample is larger than the coherently illuminated area. This problem can be overcome by the technique of X-ray ptychography, in which the incident beam is scanned over

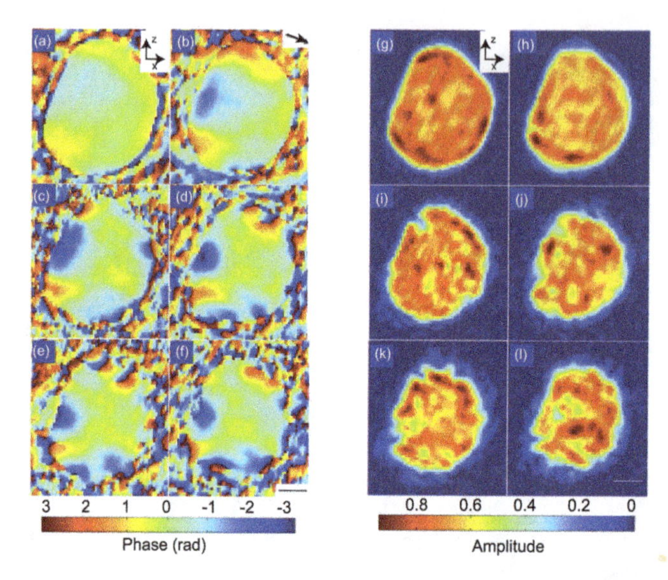

Figure 5.18. Reconstructed phase and amplitude images from the BCXDI diffraction patterns of figure 5.17. Phase reconstructions (a)–(f) and amplitude reconstructions (g)–(l) correspond to the time intervals after Cu deposition of 0 to 10 hr. Reproduced from Xiong *et al* (2014). CC BY 4.0.

the sample, recording CXDI diffraction patterns at each point. By ensuring that there is sufficient overlap of these illuminated areas the complete data set over-determines the problem and ensures that phase retrieval can be achieved reliably. With improvements in synchrotron radiation emittance, and the availability of FEL sources with much improved coherence, it seems likely that these lens-less imaging techniques will become increasingly important to understand nanocrystalline properties. Miao *et al* (2015) have reviewed a number of variations and developments of the general CXDI methods that have been explored.

References

Amati M, Susi T and Jovičecić-Klug P 2023 *J. Electron Spectrosc. Relat. Phenom.* **265** 147336

Bauer E 2012a A brief history of PEEM *J. Electron. Spectros. Rel. Phenom.* **185** 314

Bauer E 2012b LEEM and UHV-PEEM: a retrospective *Ultramicroscopy* **119** 18

Engel W, Kordesch M E, Rottermund H H, Kubala S and von Oertzen A 1991 A UHV-compatible photoelectron emission microscope for applications in surface science *Ultramicroscopy* **36** 148

Ghidini M, Zhu B and Mansell R *et al* 2018 Voltage control of magnetic single domains in Ni discs on ferroelectric $BaTiO_3$ *J. Phys. D: Appl. Phys.* **51** 224007

Ghidini M, Mansell R and Pellicelli R *et al* 2020 Voltage-driven annihilation and creation of magnetic vortices in Ni discs *Nanoscale* **12** 5652

Ghidini M, Maccherozzi F, Dhesi S S and Mathur N D 2022 XPEEM and MFM imaging of ferroic materials *Adv. Elect. Mater.* **8** 2200162

Hammond C and Imam M A 1991 Photoemission electron microscopy in metallurgical research: applications using the Balzers KE3 metioscope *Ultramicroscopy* **36** 173

Hellmann S, Sohrt C and Beye M *et al* 2012 Time-resolved X-ray photoelectron spectroscopy at FLASH *New J. Phys.* **14** 013062

Johansson L I, Armiento R and Avila J *et al* 2014 Multiple π-bands and Bernal stacking of multilayer graphene on C-face SiC, revealed by nano-angle resolved photoemission *Sci. Rep.* **4** 4157

Liang M, Lin W and Zhao Q *et al* 2023 Spatially resolved local electronic properties of two-dimensional lead halide perovskite single crystals studied by X-ray photoemission electron microscopy *Solar RRL.* **7** 2200795

Menteş T O and Locatelli A 2012 Angle-resolved X-ray photoemission electron microscopy *J. Electron Spectros. Rel. Phenom.* **185** 323

Miao J, Ishikawa T, Robinson I K and Murnane M M 2015 Beyond crystallography: diffractive imaging using coherent X-ray light sources *Science* **348** 530–5

Ning Y, Fu Q and Li Y *et al* 2019 A near ambient pressure photoemission electron microscope (NAP-PEEM) *Ultramicroscopy* **200** 105

Niu Y, Vinogradov N and Preobrajenski A *et al* 2023 MAXPEEM: a spectromicroscopy beamline at MAX IV laboratory *J. Synchrotron Rad.* **30** 468

Schmidt T, Sala A, Marchetto H, Umbach E and Freund H-J 2013 First experimental proof for aberration correction in XPEEM: resolution, transmission enhancement, and limitation by space charge effects *Ultramicroscopy* **126** 23

Schneider C M and Schönhense G 2002 Investigating surface magnetism by means of photo-excitation electron emission microscopy *Rep. Prog. Phys.* **65** 1785

Schroer C G, Boye P and Feldkamp J M *et al* 2008 Coherent X-ray diffraction imaging with nanofocused illumination *Phys. Rev. Lett.* **101** 090801

Stöhr J, Scholl A and Regan T J *et al* 1999 Images of the antiferromagnetic structure of a NiO (100) surface by means of X-ray magnetic linear dichroism spectromicroscopy *Phys. Rev. Lett.* **83** 1862–5

Takayama Y, Takami Y, Fukuda K, Miyagawaa T and Kagoshima Y 2018 Atmospheric coherent X-ray diffraction imaging for *in situ* structural analysis at SPring-8 Hyogo beamline BL24XU *J. Synch. Rad.* **25** 1229–37

Tonner B P and Harp G R 1988 Photoelectron microscopy with synchrotron radiation *Rev. Sci. Instrum.* **59** 853

Tonner B P and Harp G R 1989 Photoyield spectromicroscopy of silicon surfaces using monochromatic synchrotron radiation *J. Vac. Sci. Technol.* A **7** 1

Tromp R M, Fujikawa Y and Hannon J B *et al* 2009 A simple energy filter for low energy electron microscopy/photoelectron emission microscopy instruments *J. Phys. Condens. Matter* **21** 314007

Tromp R M, Hannon J B and Ellis A W *et al* 2010 A new aberration-corrected, energy-filtered LEEM/PEEM instrument. I. Principles and design *Ultramicroscopy* **110** 852

Tromp R M, Hannon J B, Wan W, Berghaus A and Schaff O 2013 A new aberration-corrected, energy-filtered LEEM/PEEM instrument II. Operation and results *Ultramicroscopy* **127** 25

Winkler P, Zeininger J and Suchorski Y *et al* 2021 How the anisotropy of surface oxide formation influences the transient activity of a surface reaction *Nat. Commun.* **12** 69

Xiong G, Clark J N, Nicklin C, Rawle J and Robinson I K 2014 Atomic diffusion within individual gold nanocrystal *Sci. Rep.* **4** 6765

Printed in the USA
CPSIA information can be obtained
at www.ICGtesting.com
LVHW082342150324
774517LV00005B/719